Stem Cells

Innovation, Benefits and Risk in Biotechnology Set

coordinated by
Nicole Arrighi and Christine Risso

Stem Cells

Therapeutic Innovations Under Control

Nicole Arrighi

iSTE PRESS

ELSEVIER

First published 2018 in Great Britain and the United States by ISTE Press Ltd and Elsevier Ltd

ISTE Press Ltd
27-37 St George's Road
London SW19 4EU
UK

www.iste.co.uk

Elsevier Ltd
The Boulevard, Langford Lane
Kidlington, Oxford, OX5 1GB
UK

www.elsevier.com

British Library Cataloguing-in-Publication Data
A CIP record for this book is available from the British Library
Library of Congress Cataloging in Publication Data
A catalog record for this book is available from the Library of Congress
ISBN 978-1-78548-254-0

Contents

Acknowledgments

I wish to thank all those who have accompanied me in this work.

Angèle Chopard, Professor at the University of Montpellier, "UMR 866 Muscular Dynamic and Metabolism" team, whom I thank for her friendly energy and unwavering support.

Christian Dani, Research Director, Head of the "Stem Cells and Differentiation" team at the Institute of Biology Valrose (CNRS-UMR 7277 INSERM U1091), University of Nice, whom I thank for his sound advice.

Pascal Peraldi, INSERM Researcher for the "Stem Cells and Differentiation" team at the Institute of Biology Valrose (CNRS UMR 7277 INSERM U1091), University of Nice, whom I thank for his amicable contribution.

Introduction

> *"Biomedical research, in its broadest sense, that is to say, widely crossing over into human sciences, can permeate, influence, guide – in a word, humanize – all great scientific advances and prepare humankind for its future".*
>
> (Simone Veil, French Health Minister, 1975)

Stem cells are the original cells of any organism. They are those of the embryo in the early stages of its development. Starting from approximately 30 stem cells contained in the embryo, the development program of the whole organism occurs. From identical embryonic cells, they progressively engage towards specialization pathways, at the origin of tissues and functional organs. Historically, the determinism of the creation of the human being, which is so complex, from a simple embryo, has been investigated by researchers exploring developmental biology. Embryonic stem cells were discovered after the stem cells found in the adult organism. The latter have been isolated in tissues such as skin or muscle, where they play a key role in their constitution and their renewal.

Since these stem cells have been maintained in proliferation *in vitro*, a new benefit has emerged. They can multiply to infinity and produce all of the cells of the human body. They represent a tremendous hope in the field of health. These stem cells create a technological platform that provides solutions to medicine and research. Are they able to regenerate a damaged tissue? Are they able to replace sick or aging cells? Are they a weapon against age-related disorders?

The questions are challenging and complex. Based on knowledge of cell development, these new biotechnologies call on mechanisms where the cell orchestrates the organism's repair by specializing in all of a tissue's functional cells. The coupling of stem cells with biomaterials offers unhoped-for prospects. This breakthrough in regenerative medicine has resulted in the invention of tissue engineering and created "living" medicines. These new medicines concern us all. It should be possible for all individuals to form their own opinion on the benefits and risks of these innovative therapies. The ability to estimate the safety of a treatment should be accessible to all. In order to make personal decisions, often in a hurry, an in-depth view is required, a clear and comprehensible summary of the most innovative research, which is offered in this book.

The aim of this book is to enter the world of stem cells and address the subject in its entirety. The stem cell approach is opposed to a compartmentalized vision, because they form a part of the *complexity of the living*. Beyond a scientific history, stem cells surpass the stakes of medicine. Handling and using stem cells has an impact on the living. What are the risks in terms of safety and toxicity for living beings? What does the law say? Is the legal framework the same in all countries? Do "cellular" medicines exist on the international market? From tissue regeneration to the era of augmented man, the ethical choices must be considered. While stem cells are at the core of therapeutic innovation, they bring a whole network of teams of doctors, researchers and companies into competition in a race for performance. With the acceleration of science, is this innovation really under control?

This book proposes to offer part of the response to these many questions, structured in five chapters. The starting point of this book describes the key stem cell discoveries. Based on observation, it provides pathways of scientific reasoning from discovery up to the monitoring of scientific progress in this complex subject.

First, the unique characteristics and the different types of stem cells will be defined (Chapter 1). After this classification, the use of stem cells as a new tool for studying diseases will be discussed. Decrypting the mechanism of a pathology and imagining treatment paths through the screening of pharmaceutical molecules: these are the responses provided by stem cells thanks to this experimental platform necessary to medical research (Chapter 2). Next, Chapter 3 will present the prospects of cell therapy through the transplantation of stem cells in an injured tissue or organ and, beyond the transplant, the design of tissue reconstruction, combining a whole microenvironment with stem cells, which is the starting point of regenerative medicine, explored in Chapter 4. This rapid evolution of fundamental research and translational research has raised bioethical and regulatory questions that will be developed in Chapter 5.

1

Definition and Classification of Stem Cells

> "*I was captivated by the idea of understanding and controlling the pluripotent state*".
>
> (Austin Smith, Director of the *Wellcome Trust Centre for Stem Cell Research*, Cambridge)

Stem cells are the original cells of the human being. With infinite multiplication potential and the ability to differentiate into any type of cell of the organism, stem cells are the foundations of tissues, organs and the entire organism. Throughout life, they regenerate and repair damaged tissues following injury or illness. Amplified from a sample taken from the patient or genetically modified, these cells grafted onto a support guarantee an optimal anchoring and a rapid proliferation. This living graft, transplanted onto the damaged tissue area, is capable of introducing cells that restore its function. These stem cells form the basis of the tissue repair processes. But what is a stem cell? Where does it come from? Where in the organism is it located? What is its role?

The aim of this chapter is to define the unique properties of stem cells (SCs), locate their source and understand their development. In the human organism, several categories of SCs exist; this chapter is organized into three parts corresponding to the three categories of stem cells.

First, the stem cells present in the various tissues of the human organism will be addressed. These are called adult stem cells and are located in specific tissues or organs, part of the adult organism. Bone-marrow-, muscle-, skin- and adipose tissue-derived SCs as well as their physiological functions and the molecular markers that characterize them are thus presented. All SCs of the tissues or organs of the organism have a common origin: they come from SCs of the embryo. Each of the embryonic SCs can give rise to an entire organism. Second, in section 1.3, we will examine these embryonic SCs, situating the early stages of the embryo's development and their role in the first weeks after fertilization. Finally, section 1.4 presents the discovery that has given a boost to regenerative medicine: the reprogramming of somatic cells into induced pluripotent cells. As if they were going back in time to the embryonic stage, these stem cells again become pluripotent and cover potentialities close to those of embryonic SCs. Sampled from the adult individual and not from the embryo, they pave the way for tissue regeneration therapies, which could have been restrained by the ethical constraints linked to embryonic SCs.

1.1. Two characteristics specific to stem cells

SCs are original cells that are considered to be the progenitors of more than 200 types of cells in the human body. They are able to divide and produce other cells that can become highly specialized. Cells are said to be SCs if they possess the following two characteristic properties: *self-renewal* and *pluripotency*. Self-renewal is the ability to multiply infinitely, by simple division. Pluripotency means that the cell is capable of producing all of the organism's cell types. These properties can be demonstrated *in vitro* where the progeny of the SCs is characterized by the expression of specific genes of a cell *lineage*. This process is called *differentiation*.

Thanks to the aforementioned two properties, SCs continually reconstitute the organism's stock of cells. They assure the renewal of the cells, which have different lifecycles. While the intestine cell is renewed after 5 days and that of the retina after 10 days, other cells live longer, such as the red blood cell, which is renewed after 120 days, or the liver cell, which lives for approximately 400 days. The subject of cancerous cells will intentionally not be discussed in this book. Although they present an infinite multiplication, they do not, however, have the capacity to generate all of the tissues of the organism.

1.1.1. *An infinite multiplication capacity: self-renewal*

The future for SCs – self-renewal or differentiation – is determined through the process of cell division. SCs produce undifferentiated SCs or specialize into progenitor cells. This process is regulated by symmetric and asymmetric divisions. It depends on cell-intrinsic and cell-extrinsic factors, such as microenvironmental *stimuli*. It is then possible for the signals from a specific microenvironment to be able to control SC dynamics. For example, SCs with a wide differentiation potential exist in bone marrow. SCs are also located in the tissues or in *niches* that provide a specific cellular environment indispensable for unlimited renewal.

Symmetric division gives rise to two identical daughter cells possessing the properties of the SCs. Asymmetric division produces an SC and a progenitor cell with limited self-renewal potential. The *progenitors* can perpetuate several cycles of cell division before finally differentiating into a specialized cell. The molecular differences between *symmetric and asymmetric division* would be based on a differential selection of membrane proteins (receptors) between the daughter cells. The SC continuously generates daughter cells that are invariably identical to it and, in parallel, generates other daughter cells that have different, more restricted properties.

The capacities for *self-renewal and specialization are inverse*. The SC is not specialized and possesses maximum self-renewal capabilities. Through the process of differentiation, it produces the *progenitors* and then the *precursors*, which is the last step toward the production of the *specialized cell*. This differentiated cell has lost its self-renewal capability; however, its specificity is maximum.

SCs can differentiate into several cell types, and this broad range of differentiation orientations is called SC *plasticity*. They have the ability to move from one type of differentiated cell to another, that is, *transdifferentiation*.

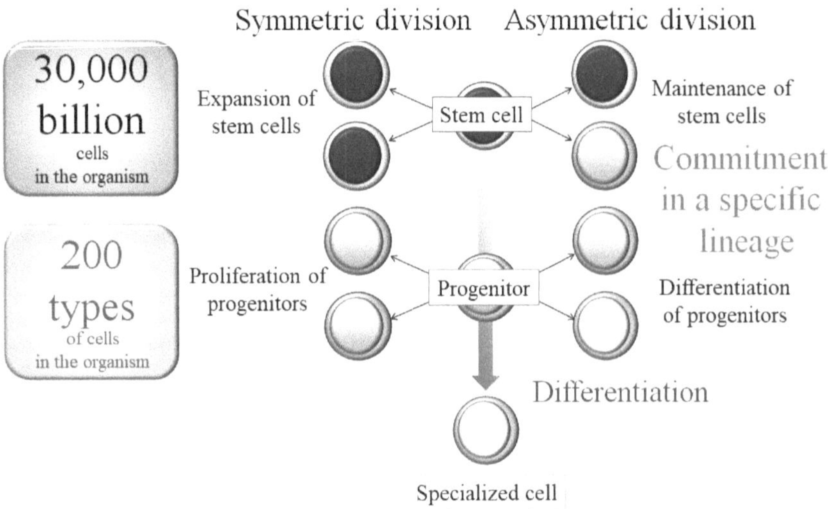

Figure 1.1. *Self-renewal of SCs and progenitor cells. Symmetric division produces two identical cells, whereas asymmetric division leads to two different cells. For a color version of this figure, see www.iste.co.uk/arrighi/stemcells.zip*

1.1.2. *A specialization ability: differentiation*

The organism's original SCs are present in the fertilized egg, also known as the *zygote*. The zygote is a cell resulting from the fusion of two gametes. At this first stage in development, the fertilized egg possesses all differentiation capacities. The cells, which compose it, are called *totipotent*, from the Latin *totus*, meaning "everything". They are fit for everything in the sense that they have the ability to become a complete individual. They can form an embryo, as well as the placenta. The egg divides into two, four and then eight cells. After 4 days, the egg resembles a blackberry and hence this stage is named *morula*, meaning "blackberry" in Latin.

After 5–6 days, the embryo reaches the stage of blastocyst, from the Greek *blastos*, meaning 'bud". It takes the form of a hollow ball and consists of approximately 200 cells. The peripheral cells, or trophectoderm, will form the placenta, necessary for embryonic implantation. Inside this ball are approximately 30 cells, from which more than 200 different cell types of the organism can develop. This internal cellular mass contains SCs that are *pluripotent*, from the Latin *pluris*, which means "the most", in the sense of a larger quantity. They can diversify into any tissue of the organism, but they are unable to produce a complete individual. Pluripotent cells produce three embryonic layers, namely the endoderm, the mesoderm and the ectoderm. Each layer generates determined cells for each organ.

The embryonic layers contain cells that are *multipotent*, from the Latin prefix *multi*, which means "many". They are capable of diversifying into a limited number of cell types with the same embryonic origin.

Each layer generates determined cells for each organ or line, known as *unipotent* SCs. They undergo a unique lineage that enables the production of a specific cell type.

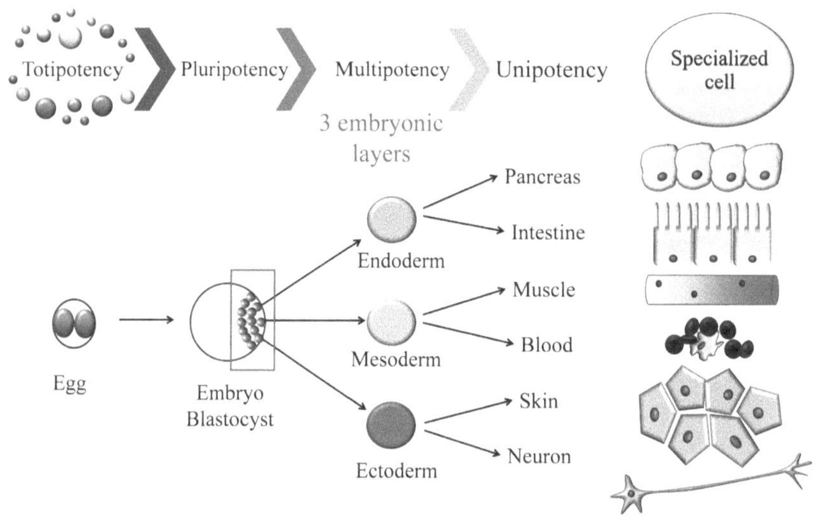

Figure 1.2. *The fate of SCs from the totipotency of the egg to the pluripotent cells of the embryo, then multipotent cells of the embryonic layers and finally specific cells of each organ of the human body. For a color version of this figure, see www.iste.co.uk/arrighi/stemcells.zip*

The SCs present in the human body are multipotent or unipotent. They do not exhibit any specialized physiological property. They are categorized into the following three large families, whose characteristics and locations are different:

– *Embryonic* SCs, or ES (embryonic stem) cells, are derived from the human embryo a few days after fertilization. The embryo is in the early stages of its development; it is not yet implanted, hence the term "preimplantation embryo". At this stage, it contains a compact mass of cells, known as *inner cell mass*, which are the pluripotent embryonic SCs.

– The *adult* or *tissue-specific* SCs are located in the tissues and organs of the human body, yet in minimal amounts. They are responsible for tissue renewal, differentiating into any specialized cell type of the tissue or organ. Their primary role is to maintain and repair the tissue where they are located, such as, for example, skin or muscle SCs.

– *Induced pluripotent* SCs (or iPS cells), which have been discovered very recently, are obtained in the laboratory by genetically reprogramming specialized somatic cells into cells with self-renewal and pluripotency capabilities, two functional properties that make them closely resemble human embryonic SCs.

In the following sections, we will detail the three major families of SCs, examining, to begin with, adult SCs, then embryonic SCs and finally induced pluripotent SCs, identifying their discovery and specific properties. These SCs resemble a scientific platform that offers immense hope and brings solutions to the needs of medicine.

1.2. Adult stem cells or specific stem cells of a tissue

Adult SCs are cells of the organism that divide, replace dead cells and regenerate damaged tissues. They are somatic cells, i.e. of the body (*soma* in Greek). They are located within a precise tissue or organ. They possess two properties common to all SCs: (1) the ability to renew to infinity, although less so than embryonic SCs, driving several division cycles while remaining in undifferentiated state and (2) pluripotency, that is, the ability to generate a progeny of different cell types.

Among these adult SCs are the *tissue-specific SCs* such as bone-marrow-, skin-, muscle- and adipose tissue-derived SCs, as well as those of the hematopoietic system, although this is not, strictly speaking, a tissue.

1.2.1. *Bone-marrow stem cells*

Bone marrow is a tissue that is found inside bones, which constitutes the main source of SCs. It produces hematopoietic SCs and mesenchymal SCs.

Hematopoietic SCs, on the one hand, are at the origin of all blood cells: first red blood cells that transport oxygen; then megakaryocytes, producers of platelets, which interact with the blood vessels and coagulation factors; and finally the cells of the immune system, which act against microbe attacks.

On the other hand, mesenchymal SCs, discovered very recently in bone marrow, can produce cartilage, bone and fat cells. Although located in the same tissue niche, these two types of SC exhibit different properties and functions.

1.2.1.1. *Hematopoietic stem cells*

By definition, hematopoiesis (from the Greek *haímatos*, meaning "blood", and *poiêsis*, meaning "creation") is the set of mechanisms involved in the production of blood cells from *hematopoietic SCs*, also called *blood SCs*.

Blood SCs were discovered by Canadian researchers James Till and Ernest McCulloch at the Ontario Cancer Institute (Toronto, Canada) in 1961. They observed that the injection of a reduced number of bone-marrow cells into radiated mouse spleen causes the appearance of nodules consisting of several hematopoietic lineages within 8 days. They published, in the review *Radiation Research*, the first quantification of hematopoietic SCs [TIL 61]. They discovered that these bone-marrow cells possess two particular abilities. These cells have the potential to renew and differentiate in order to produce all of the cell lineages of the blood system. Till and McCulloch attributed to these cells the name "stem" cells as they are at the summit of the hierarchy of the hematopoietic system and make cellular transplantation possible. Their work was a scientific breakthrough that launched research into SCs.

Based on the role of cells and on their quantification rather than morphological observations, their study revealed the existence of a cell capable of reproducing and producing descendants acquiring the particular characteristics of specialized blood cells. They highlighted self-renewable cells in mouse bone marrow. This first description of an SC formed the subject of a publication in the review *Nature* in 1963 [BEC 63].

In adults, hematopoietic SCs generate one billion red blood cells and 100 million white blood cells per hour, as well as platelets and other mature blood cells. Hematopoietic SCs must also self-renew and respond to external *stimuli* so as to maintain hematopoietic homeostasis. SCs that are hematopoietic are represented at the top of the lineage tree, each branch of which limits the differentiation capabilities a little more.

This hematopoietic system consists of a hierarchy of cells that develop from stem state to differentiated state through approximately 10 blood cell lineages. While the hematopoietic SCs are few in number, have a long lifecycle and are infinitely prolific, other cell progenitors lose their self-renewal and pluripotency capabilities before becoming mature blood cells. In the conventional model of hematopoiesis, a major bifurcation separates the myeloid and lymphoid branches. The myeloid progenitors then undergo a second bifurcation between progenitors of macrophage-granulocytes and those of erythrocyte-megakaryocytes (large cells at the origin of platelets). These bipotent progenitors produce unipotent progenitors that ultimately lead to all progeny.

During fetal development, the bone marrow produces leukocytes, platelets and red blood cells until birth, where it remains the main site of hematopoietic production. In adults, the bone marrow of the vertebrae, ribs, skull, pelvis and femur ensures the renewal of blood lines. Residing in this specialized microenvironment, called a *niche*, hematopoietic SCs will produce specific progenitors, which will feed the blood lineages throughout life. An understanding of the signals that the SCs receive in this microenvironment as they multiply and differentiate is essential for the development of therapeutic applications. The old representation of the cellular niche perceived as a static environment, defined in a precise anatomical compartment consisting of a predominant cell type, is investigated. Henceforth, this niche is increasingly conceived as an anatomical fluid in a *dynamic environment* that integrates physical or cellular messages intended for SCs.

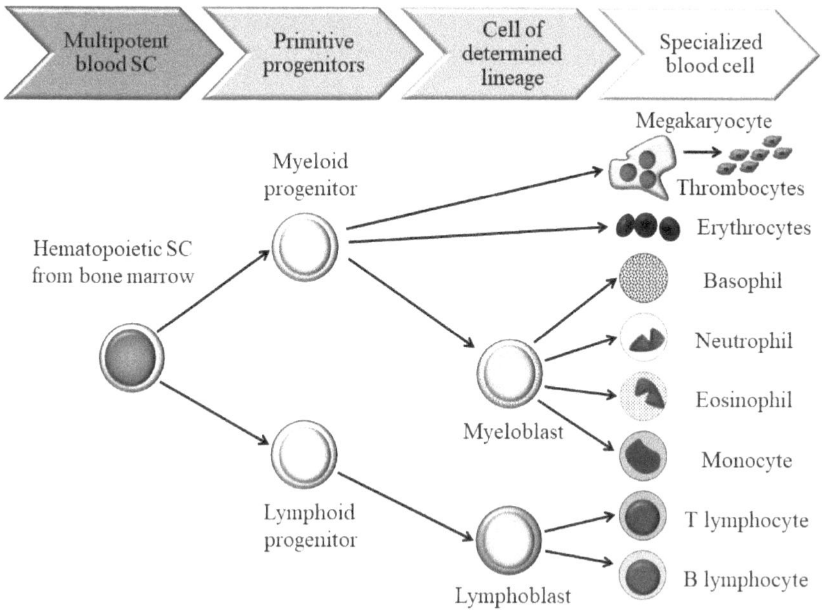

Figure 1.3. *Hierarchy of hematopoiesis in bone marrow in adult humans. Hematopoietic SCs evolve from stem state to differentiated state, through approximately 10 blood cell lineages. For a color version of this figure, see www.iste.co.uk/arrighi/stemcells.zip*

1.2.1.2. *Mesenchymal stem cells*

Bone marrow contains hematopoietic SCs and non-hematopoietic SCs, which are called mesenchymal SCs. This term refers to the *mesenchyme*, an organized embryonic tissue that produces the connective tissues of smooth muscles, cardiac muscle and vessels. By extension, in adults, mesenchymal tissue refers to the connective tissue.

Mesenchymal SCs represent a heterogeneous population that has no common identity, no locating unit and no common future. They are *multipotent* and able to differentiate into cartilage cells (chondrocytes), bone cells (osteoblasts) and fat cells (adipocytes). Discovered in bone marrow, they are found in other tissues such as umbilical cord blood, adipose tissue and muscles. The only existing difference between mesenchymal SCs derived from bone marrow and adipose tissue is based on the maintenance role of the supporting tissues, in contrast to cord blood where this function is absent. This property of reconstitution of the tissue compartments makes it the major asset of these cells in cell therapy and in regenerative medicine.

Mesenchymal SCs that are mesodermal in origin were described for the first time by Alexander Friedenstein *et al.* as non-hematopoietic, bone-marrow-adherent cells that can form colonies identical to those of fibroblasts [FRI 76]. As they have the ability to differentiate into bone and cartilage, one of the clinical applications is to combine them with biomaterials in the repair of long bone fractures. Since the demonstration of their *immunosuppression* properties, research into mesenchymal SCs has accelerated and cell therapy clinical trials are increasing in number. Although they present definite differences, they have several criteria in common:

– adhesion to the plastic support;

– the ability to form typical fibroblast colonies;

– their expression of surface markers CD90, CD73 and CD105 and the fact that they do not express the hematopoietic markers CD45, CD34 and HLA-DR (histocompatibility complex gene);

– their differentiation into osteoblast, chondrocyte and adipocyte.

These main characteristics apply to mesenchymal SCs from bone marrow, but some differences may occur, depending on the origin of the cells. Sampled under local anesthesia, easy to multiply *in vitro*, dividing rapidly and capable of forming cartilage, bones and other cell types, mesenchymal SCs are the main adult SCs used in clinical trials because of their immunomodulatory properties and differentiation potential.

1.2.2. *Umbilical cord blood stem cells*

The characterization of hematopoietic SCs took decades. The identification of bone marrow as the source of these SCs intensified research and led to the discovery of another source of hematopoietic SCs in umbilical cord blood. After birth, cord blood is sampled from the umbilical cord and placenta. It is rich in hematopoietic SCs and hematopoietic progenitors. Compared to bone marrow, it comprises cells that are more immature from an immunological viewpoint, which enables a lesser degree of compatibility to be required between the donor and the recipient. It presents a lower incidence of infectious agents and transplant rejection.

The use of umbilical cord blood as a source of hematopoietic SCs was proposed in 1982 by Hal Broxmeyer to Edward A. Boyse and Judith Bard, who founded the company Biocyte. In 1988, the first ever transplant of SCs from umbilical cord blood was performed by Professor Eliane Gluckman's team on a five-year-old child with Fanconi anemia [GLU 89]. Since then, the efficiency of transplants of SCs derived from umbilical cord blood has been extensively demonstrated for malignant and non-malignant diseases, severe and chronic leukemias, lymphomas, myelodysplastic syndromes, thalassemia, Fanconi anemia and other bone-marrow deficiencies. Since the first transplant of SCs from umbilical cord blood, more than 20,000 transplants have taken place worldwide. Since then, cord blood banks have been established worldwide. This cell therapy of human SCs relies on the functional characteristics of hematopoietic SCs isolated from peripheral blood, bone marrow or umbilical cord blood and requires the immunological barriers to be crossed in order to support the transplant and minimize the risks of rejection.

1.2.3. *Intestinal stem cells*

The human intestinal epithelium consists of villi, which are folds of the intestinal mucosa, and *crypts composed of different types of cells,* including intestinal SCs. They have served as a *model in the study of adult SC biology* because they present two advantages: they are easily accessible and ensure a permanent and extremely rapid renewal of the cells of the surface of the intestine, whose lifespan is approximately 5 days.

Intestinal SCs are maintained throughout the life of the human being. They are multipotent and generate all cell types of the intestinal lineage, comprising enterocytes, goblet cells, endocrine cells and Paneth cells. All of these cells are distributed *across the surface of the intestine*, according to a precise topography. The proliferation compartment is located *in the crypt*, while the differentiation compartment is situated in the villus. *The more the cells are specialized*, the more they migrate from the bottom of the crypt to the lumen of the intestine. *Each cell type* assures a distinct physiological function, including absorption and the endocrine/exocrine functions [CHE 74]:

– the *enterocytes* are the most abundant cells of the small intestine and are responsible for the *absorption* of nutrients and the secretion of hydrolytic enzymes in the lumen of the intestine;

– the *goblet cells* produce mucus, which protects the intestinal mucosa;

– the *enteroendocrine cells* produce *hormones* (serotonin and secretin);

– the *Paneth cells* located at the bottom of the crypts secrete *antimicrobial* agents (lysozyme), which play the role of defending the intestinal mucosal barrier.

Thanks to molecular genetics techniques, Hans Clevers' group, based at the Hubrecht Institute in Utrecht, the Netherlands, confirmed the works of C.P. Leblond, cited above [BAR 07]. H. Clevers proved that SCs located at the bottom of the crypts are at the origin of all of the intestine's differentiated cells. In the absence of a specific marker of intestinal SCs, Clevers' works explored the role of signaling pathways in the transformation of the intestinal epithelium. Activation of the signaling pathway of the Wnt/β-catenin/Tcf4 pathway is an early stage that is essential for the renewal of intestinal tissue, which allows the intestinal SCs to be traced during their differentiation from the *genetic signature of the Wnt pathway*. In addition to understanding the renewal process of adult SCs, Clevers' works enabled exploration of the maintenance of intestinal SCs under both normal and pathological conditions.

H. Clevers' group identified a specific marker of SCs in the small intestine and colon. The gene Lgr5 (leucine-rich repeat-containing G-protein-coupled receptor 5, also called "Gpr49") was isolated from a set of target genes of the Wnt pathway. The cells that express this gene are located at the bottom of the crypt of the intestine and generate all of the intestinal lineages. Lgr5 is expressed in rare cells of other tissues [BAR 07]. After identifying the intestinal SCs, H. Clevers introduced the mutation of the gene, *APC* (*adenomatous polyposis coli*), associated with intestinal tumorigenesis. Adenomas and adenocarcinomas have appeared in Lgr5 + cells, while progenitors remain unscathed, proving that the crypt SCs are at the origin of intestinal cancer [BAR 09].

The current focus of the works of H. Clevers is to produce Lgr5 + human intestinal SC culture systems. These cultures are genetically and phenotypically stable and have been transposed to the stomach or liver cells. Miniatures of human organs, or "organoids", were presented the European Inventor Award 2017, to H. Clever in the Research category.

1.2.4. *Neural stem cells*

For approximately the last 20 years, neuroscience has been exploring the functions of neural cells. The identification of progenitor cells and the lineages that they produce are the keystone of the immense diversity of the cell types produced in the brain. Neural SCs and neural progenitors are responsible for neurogenesis in embryos and in adults. During the development of the embryo, the neural SCs are located in the neural crest and cortex. In the adult brain, neurogenesis occurs in the subventricular zone (SVZ) of the lateral ventricles and the subgranular zone (SGZ) of the hippocampus near the serrated or dentate gyrus (circonvolution).

Neural SCs produce new *neurons* involved in learning and memory; this is the locus of *adult neurogenesis*, as opposed to embryonic neurogenesis.

Neural SCs in the subventricular zone are capable of generating glial cells – *astrocytes*, *oligodendrocytes* and oligodendrocyte progenitors – that reside in all regions of the central nervous system and continue to generate new myelinated oligodendrocytes all throughout life. The proliferative capacities of these populations decrease with age and are correlated with the decline in cognitive performance. However, the fact that a reservoir of quiescent precursors can be activated in the neurogenic niches widens the scope of possible therapeutic applications.

Neural SCs result from the transformation of radial glial cells during the neonatal period. They self-renew and generate neurons and glial cells *in vivo*. The presence of neural SCs resembling radial glial cells has been demonstrated in the hippocampus, which would enable the generation of neurons, astrocytes and oligodendrocytes. Neurons and glial cells express specific markers for specific cell types over time, some examples of which are as follows [ZHA 15]:

– transcription factors (Sox2, Pax6, Mash1 and NeuroD);

– enzymes (glutamine synthase);

– cytoskeletal proteins (nestin);

– membrane receptors (CD133) that enable the identification of subpopulations of neural SCs.

The generation of new neurons from neural SCs throughout life is a process that attracts research attention due to its authenticity and restoration potential. In the serrated gyrus of the hippocampus, SCs generate new granulated cells and astrocytes by a complex, plastic mechanism that can be regulated by endogenous and exogenous factors. This population of neural SCs is heterogeneous by its morphology and its potential for differentiation.

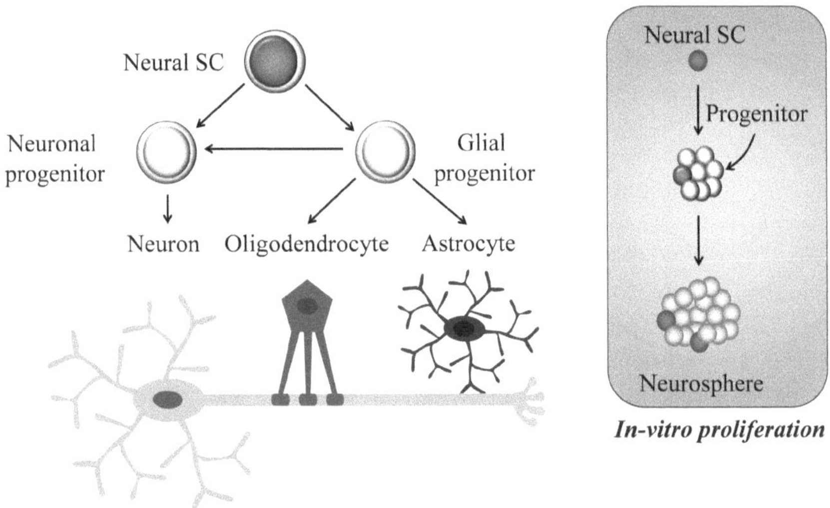

Figure 1.4. *Differentiation of neural SCs and neural progenitors in the adult human brain. Inset:* In vitro *proliferation is manifested by an aggregated cell structure, the neurosphere. For a color version of this figure, see www.iste.co.uk/arrighi/stemcells.zip*

1.2.5. *Adipose tissue-derived stem cells*

Adipose tissue is a specialized tissue, composed of fat cells, or *adipocytes*. Two types of adipose tissue exist, whose adipocytes present morphological and functional differences:

– *White adipose tissue* (WAT), the most abundant of the two, is composed of unilocular adipocytes, consisting of a single lipid vacuole swelling the cytoplasm and displacing the nucleus peripherally against the plasma membrane, hence the term unilocular adipose tissue. Located in several compartments of the human organism, mostly in the subcutaneous and visceral adipose tissues, white adipose tissue stores energy in the form of triglycerides, which are synthesized and then hydrolyzed according to the needs of the organism. In skeletal muscle, for example, a population of non-myogenic adipose progenitors has been characterized: it differentiates into perfectly functional adipocytes, presenting properties similar to adipocytes derived from SCs of subcutaneous adipose tissue [ARR 15].

– *Brown adipose tissue* (BAT), known for its thermogenic properties, produces heat by the uncoupling protein (Ucp1), which uses the mitochondrial proton gradient of oxidative phosphorylation to generate thermal energy, rather than producing energy-rich substrates such as ATP. It is activated mainly by the cold. Unlike white adipocytes, brown adipocytes are multilocular. They contain several lipid vacuoles and numerous mitochondria. In adult humans, brown adipose tissue is limited, whereas it is more abundant in newborns.

White and brown adipose tissues have such different properties and locations that the question of the origin of these tissues arises. The review by Billon and Dani traces the developmental origin of adipocytes [BIL 12]. The authors identify adipocyte progenitors of the neural crest as a new source of adipocytes *in vivo* [BIL 07]. They compare their results with the knowledge acquired, describing adipose tissue as a derivative of the mesoderm, where an adipocyte progenitor (preadipocyte) would exclude all mesenchymal derivatives. While terminal differentiation of the preadipocyte to the mature adipocyte has been widely studied, the developmental origin of adipocytes during embryogenesis has been much less so and is attracting more and more interest. White and brown lineages seem to come from two very distinct origins. Further studies seem necessary in order to identify the respective progenitors in humans. This progress would contribute to an understanding of the factors influencing the functions of adipose tissue, and the treatment of obesity and metabolic diseases.

Previously incorporated into bone-marrow mesenchymal SCs, SCs generating adipocytes have since been identified in adipose tissues. Thus, the terminology has evolved toward the designation of *adipose* SCs (adipose tissue-derived stem cells, ADSC, or adipose stem cells, ASC). They are isolated after the enzymatic digestion of the *stromal vascular fraction* (SVF) of adipose tissue. From the Latin *strauma* meaning “carpet”, here “nourishing tissue” or “connective tissue”, the SVF is defined as a raw population composed of adherent cells such as fibroblasts, endothelial cells, smooth muscle cells, pericytes, immune cells and preadipocytes. Most of these cells are removed after culture; only the adherent population remains, composed of preadipocytes [WAN 16].

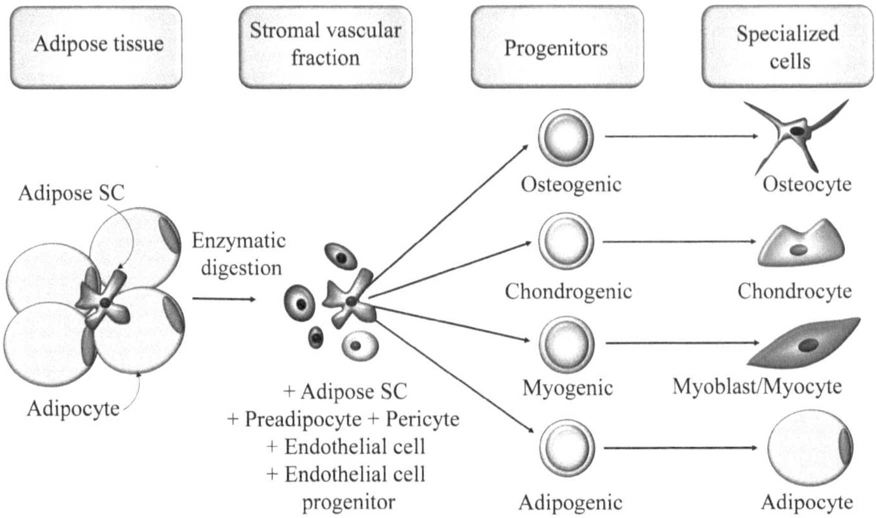

Figure 1.5. *Differentiation of adipose SCs and different lineages of progenitors. Adipose SCs are isolated from the stromal vascular fraction (SVF) of adipose tissue, a crude population composed of adherent cells such as endothelial cells, smooth muscle cells, pericytes, immune cells, preadipocytes and fibroblasts*

1.2.6. *Skeletal and cardiac muscle stem cells*

1.2.6.1. *Skeletal muscle stem cells*

Skeletal muscles are striated muscles that represent approximately 40% of the weight of the human body. Skeletal muscle is composed of contractile multinucleate muscle cells called *myofibers*. These myofibers result from the fusion of the myogenic progenitors of the mesoderm, called *myoblasts*. The number of myofibers remains constant, but each myofiber fuses with *satellite cells*, a population of muscular SCs. After an injury, the skeletal muscle undergoes a process of degeneration and regeneration based on the interactions between the satellite cells and the microenvironment of the cell niche.

The satellite cells are mononuclear SCs located outside the muscle fiber, hence the term "satellite", between the cytoplasmic membrane and the basal lamina of the muscle fiber. The population of satellite cells is quiescent. They can be activated after an injury to produce myoblasts that proliferate and then differentiate into multinucleate and contractile myotubes. This process of progenitor proliferation and then differentiation is called *myogenesis*. These satellite cells are not the only muscle SCs; they are identified by specific molecular markers, which are the PAX7 (paired box-7) transcription factors, as well as MyoD and Myogenin.

Perturbation of the niche after a muscle injury takes the satellite cells out of their quiescence. They proliferate, differentiate and fuse to existing myofibers or produce *de novo* myofibers. A family of proteins called "myogenic regulation factors" (MRFs) control the progression of the myogenic lineage: the myogenic factor-5 (Myf-5), which is the first activation marker; MyoD and Myogenin, marker of the terminal differentiation. Simultaneous activation of MyoD and Pax7 is an activation index of satellite cells. Various activators influence the lifecycle of satellite cells and their ability to participate in muscle regeneration. In humans, certain growth factors increase muscle regeneration, such as insulin-like growth factor (IGF-1), leukemia inhibitory factor (LIF), hepatocyte growth factor (HGF) and fibroblast growth factor (FGF).

The satellite cells are heterogeneous in their differentiation potential. Rudnicki's group showed that satellite cells can differentiate into several lineages of mesenchymal SCs [YIN 13]. When cultured on a Matrigel-type support, they differentiate into myocytes, adipocytes and osteocytes and resemble the mesenchymal SCs derived from bone marrow. Clonal analysis has shown that the myogenic and adipogenic satellite cells are two separate populations and both express the Pax7 myogenic marker and the adipogenic markers of the proteins, peroxisome proliferator activated receptor γ (PPARγ) and CCAAT/enhancer-binding proteins (C/EBPs). This molecular signature reflects a common developmental origin of satellite cells and adipogenic progenitors.

Myogenesis is the main fate of satellite cells *in vivo* because fibrosis and adipose infiltration are not observed in healthy, young human muscle. Recent studies showed the presence of intramuscular adipocytes and fibroblasts derived from fibroadipogenic progenitors (FAP), which reside in the muscle interstitium [ARR 15].

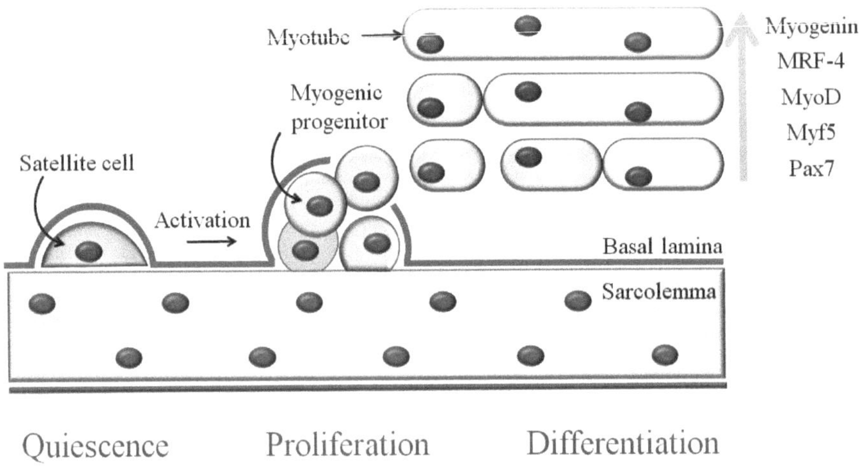

Figure 1.6. *Differentiation of muscular SCs from the satellite cell that leaves its quiescence and produces myoblasts that differentiate into polynucleate and fused muscle fibers. For a color version of this figure, see www.iste.co.uk/arrighi/stemcells.zip*

1.2.6.2. *Cardiac muscle stem cells*

The heart is a post-mitotic organ. This view has changed, however, with the identification of SCs and progenitors residing in the adult heart muscle. Cardiac progenitors form a heterogeneous group of cells distributed in the muscle tissue of the heart, or *myocardium*, including the atria, ventricles, epicardium (the outer layer of the myocardium) and pericardium (the thin membrane enveloping the myocardium). Under normal conditions, cardiac progenitors are quiescent and do not contribute to the renewal of *cardiomyocytes*, specific muscle cells of the heart, endowed with the ability to contract. After an injury or a heart attack, these progenitors are activated and differentiate into new myocytes and into vascular cells. Unlike bone-marrow SCs, where surface markers are defined, cardiac progenitors express various markers. Several subpopulations have been reported in the adult heart: c-Kit+ progenitors; cardiosphere-derived cells; epicardium-derived cells; lateral-population cells, identified by their ability to exclude Hoechst stain from the nuclei; progenitors expressing the SC antigen-1 (Sca-1^{+}); progenitors expressing Islet-1 protein (Isl-1+) and progenitors expressing PDGFR receptor (platelet-derived growth factor receptor-alpha, PDGFRα^{+}).

The first cardiac progenitors were identified in 2003 by the expression of the tyrosine kinase receptor *c-Kit* and the absence of a marker of the hematopoietic lineage. Another population of progenitors is that of *cells derived from cardiospheres*. They consist of a mixture of progenitors derived from biopsies taken from the atrium or the ventricle. In suspension culture, *in vitro*, they form aggregates of cells in clusters or cardiospheres. They are clonogenic and have a multilineage potential.

Other populations of cardiac progenitors have been identified by membrane markers (Sca-1, Abcg-2, Flk-1, CD34, CD90 and CD105) and transcription factors (Isl-1, Nkx2.5, MEF2C and GATA4) in the adult and embryonic heart. These progenitors possess the self-renewal and multipotency properties *in vivo* and *in vitro*. However, recent studies showed that Isl-1 was not a marker of endogenous cardiac CSs, but rather cells from the neural crest. In mice, recent works showed that c-Kit+ cardiac resident cells are not a source of myocytes, but rather a source of endothelial cells. In short, the exact biological role of progenitor populations in the event of cardiac injury remains unknown.

Tyl *et al.* reviewed the different types of cardiac progenitors involved in regeneration [TYL 16]. They identified a population of PDGFRα^+ cardiac progenitors expressing tyrosine kinase forming fibroblastic-type colonies in the human heart and in the mouse heart. These cells differentiate *in vitro* into vascular cells, fibroblasts and cells close to myocytes.

The use of cardiac progenitors, either by exogenous supply or from endogenous cells, has shown positive effects on regeneration following a myocardial infarction. The molecular mechanisms involved are not understood, however. In addition, the role of the cardiac fibroblast population is unknown in the regeneration of cardiomyocytes.

1.2.7. *Skin stem cells*

The skin is the largest organ of the human body, and it protects against external aggressions, infections and dehydration and provides thermoregulation and sensory perception. The skin consists of two tissues: there is the *epidermis*, a stratified, non-vascularized external epithelium, which is located above the *dermis*, a connective tissue composed mainly of dense fibrous compounds produced by fibroblasts. The epidermis is composed of two distinct layers: a locus of progressive differentiation of the *keratinocytes* migrating from the basal layer to the stratum corneum and the outermost layer. The epidermis and the hair follicle regenerate thanks to the SCs, which are quiescent and undifferentiated and ensure the maintenance of homeostasis and tissue regeneration. Several populations of SCs exist in the skin. These populations include the epidermal SCs, dermal mesenchymal SCs, hair follicle SCs, endothelial and hematopoietic SCs as well as cells derived from the neural crest, such as the melanocytes.

The epidermis is the outermost layer of the human skin. It forms a protective, water-resistant envelope composed of a stratified squamous epithelium that rests on a basal lamina. The epidermis contains four cell types, namely keratinocytes, melanocytes, Langerhans cells and Merkel cells. Keratinocytes, which produce keratin, are the predominant cells of the epidermis. In the deepest part of the epidermis, called the basal layer, are the immature keratinocytes, at the proliferative stage, responsible for *keratinopoiesis* and renewal of the epidermis, which lasts 3–5 weeks. The classical model explaining the maintenance of skin homeostasis proposes that the SCs of the basal lamina divide more slowly than daughter cells, called transit amplifying cells. After several divisions, they differentiate and transit to the upper layers. The dermis houses several functional compounds of the skin, including the vascular and lymphatic systems and epidermal appendages (hair follicles, sebaceous glands and sweat glands).

Epidermal SCs are characterized by the expression of cytokeratins KRT15 and KRT19, α6 integrin and the absence of transferrin CD71 receptor.

The *hair follicle* is a mini-organ of the skin that contains SCs from various developmental origins with various differentiation potentials. It represents a niche of multipotent epithelial SCs. Several markers have been explored in mice, but only a few are confirmed in humans. The markers CD200, KRT15 and KRT19 are positive, whereas the cells do not express CD34, NES or LHX2. More generally, these SCs have a profile of surface markers close to mesenchymal SCs that are positive for markers CD90, CD44, CD49b, CD105 and CD73 and differentiate into bone, fatty tissue, cartilage or smooth muscle. SCs of the hair follicle are activated by various aggressions and migrate toward the epidermis to participate in its regeneration. They are involved in the homeostasis of the sebaceous gland. In addition, the hair follicle contains melanocyte SCs, which are responsible for the regeneration of melanocytes, a type of pigment cell. Melanocytes produce melanin and therefore play a vital role in the pigmentation of skin and follicles.

From the dermis, *skin-derived precursors* (SKP) have been identified. They share certain characteristics with embryonic SCs of the neural crest. Multipotent, they are capable of producing *in vitro* a progeny of cells that are both neural and mesodermal. They have different properties from the other skin precursors. Transplanted in an adult mouse, these SKPs have a morphology similar to that of endogenous fibroblasts; they express the fibroblast markers of the dermis: PDGFRα, type-I collagen, vimentin, fibronectin and fibroblast-specific antigens. *In vivo*, SKPs are recruited near the fibrotic lesion and differentiate into myofibroblasts in the presence of serum.

Although universal membrane markers of the SCs of the dermis are sought, no specific marker has yet been identified. On the contrary, the SCs and the *progenitors of the human dermis* (dermal stem/progenitor cells, hDSPCs) present three specific intracellular markers:

– *SOX2*, the transcription factor, SRY, plays an essential role in maintaining the stem nature of embryonic SCs and certain adult SCs, as in the dermal papillae, which are small protuberances of the dermis, which plunge into the epidermis, leading the oxygen and nutrients to the inner layers of epidermal cells.

– *NANOG*, another marker of embryonic SCs, is also expressed in the SC populations of the dermis and bone marrow.

– *S100B*, the intracytosolic calcium-binding protein is overexpressed in the SCs of the skin.

Shim *et al.* demonstrated that the SC population and progenitors of the human dermis can be enriched by its ability to adhere to type-IV collagen [SHI 13]. They can differentiate into mesoderm cells such as adipocytes, osteoblasts, chondrocytes and ectodermal progeny, such as neuronal cells. Three-dimensional suspended cultures have shown an increase in the expression of the three specific markers, compared to a two-dimensional culture.

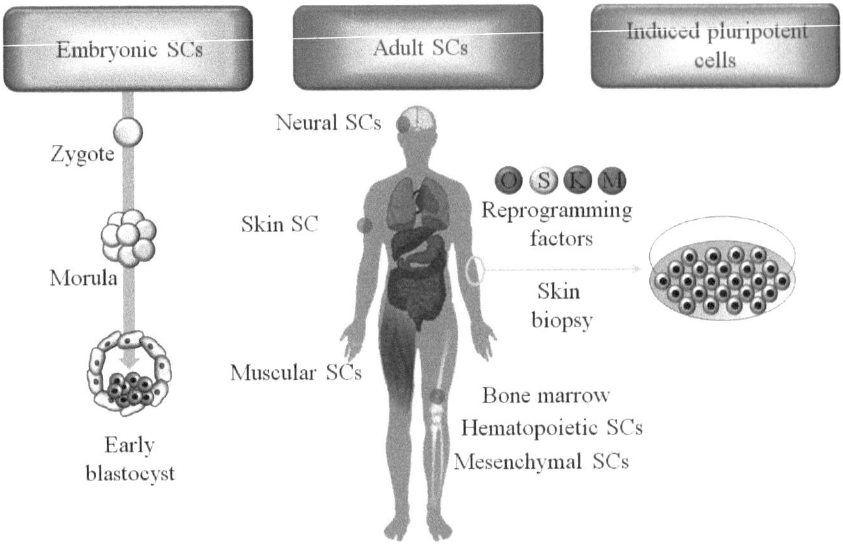

Figure 1.7. *Human SCs are divided into three families: adult SCs, embryonic SCs and induced pluripotent SCs resulting from the reprogramming of somatic cells. For a color version of this figure, see www.iste.co.uk/arrighi/stemcells.zip*

This panorama of adult SCs shows that there are as many types of SCs specific to a tissue as organs. Advanced in specialization, these progenitors are localized in a tissue and constitute its cell regeneration reservoir. At this stage of differentiation, the tissue-specific SCs cannot regenerate a different tissue than the one where they are located. To do so, it would be necessary to reprogram them and return to the original state they had in the embryo, where all directions were possible. Embryonic SCs present absolute pluripotency described in the following section.

1.3. Embryonic stem cells

The discovery of embryonic SCs has profoundly disrupted the biology of recent decades. In 1981, Martin Evans, Matthew Kaufman and Gail Martin isolated and cultured mouse *embryonic SCs* [EVA 80]. In 1998, James Alexander Thomson *et al.* found these same cells in humans [THO 98]. Since then, a whole section of research has focused on the study of embryonic SCs and has attempted to understand the origin of pluripotency. Limited by the ethical brake and the source of ESC in humans, research has invested the pluripotency mechanism on the murine model. At what moment does the embryonic SC switch from the pluripotency state to the beginning of specialization? Is its SC fate sealed?

First, the pluripotency state will be defined. We will then explore the progress made in research and the identification of different successive pluripotency states, in which different SC subpopulations of the embryo participate, such as the SCs of the epiblast, for example.

1.3.1. *Pluripotent embryonic stem cells*

Embryonic SCs (ESCs) are found in the very early embryo. At the very beginning of its growth, 5–7 days after fertilization in humans, an embryo goes through a developmental stage called the *blastocyst*, or preimplantation embryo. It is then formed of two cell types, namely the peripheral cells – the trophoblasts, which are at the origin of extra-embryonic structures such as the placenta and the umbilical cord – and the *inner mass cells* (IMC), which produce the embryo proper. The human ESC lines are derived from the *in vitro* proliferation of a fraction of the cells of the inner mass of the blastocyst. When the egg divides, it produces totipotent or pluripotent cells, which have the same future cell-production potential as the egg itself. SCs induce the formation of the three embryonic layers, namely the ectoderm, the mesoderm and the endoderm. The ectoderm will produce the skin and the nervous system. The endoderm will produce the wall of the digestive tract, and, between the two layers, the mesoderm will produce a large number of organs. Each of these compartments will have their own SCs, which will no longer be totipotent. As the layers specialize into tissues and organs, the cells become increasingly less pluripotent.

Pluripotency develops in the IMCs of the blastocyst not yet implanted in the uterus. Austin Smith's team demonstrated that the output of the pluripotency state is dictated by two transcription factors, namely Oct4 and Nanog [CHA 03]. They confer to a subpopulation of IMCs, the *epiblast*, the potential to generate the cells of the embryo. The cells of the epiblast persist briefly *in vivo*. However, they can be immortalized *in vitro* in the form of ESCs.

The *OCT4* (octamer-binding transcription factor 4) protein is used as the key marker of ESCs and for the pluripotent cells of the intact embryo. Its expression must be maintained at a critical level for the ESCs to remain undifferentiated.

The *NANOG* protein is essential for maintaining the undifferentiated state of mouse ESCs. The expression of NANOG decreases rapidly as soon as the mouse ESCs differentiate. When its level of expression is maintained by a constitutive promoter, the mouse ESCs can remain in an undifferentiated state and continue to proliferate. NANOG is also expressed in human ESCs at a lower level than OCT4.

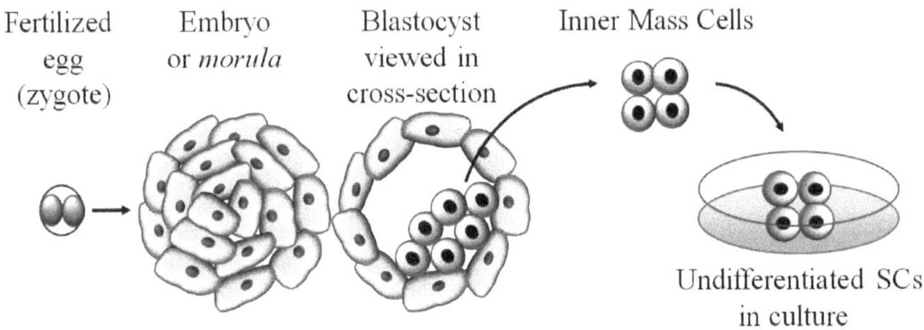

Figure 1.8. *SC culture in vitro from a fraction of the cells of the internal mass of the blastocyst*

A. Smith undertook to simplify the ESC culturing process, in order to define the essential conditions for maintaining pluripotency. During his postdoctoral studies, he was struck by an observation by a doctoral student who had noted that co-culture of SCs with a preculture of feeder cells could be replaced by conditioned medium. He concluded that a growth factor can block differentiation and is oriented toward IGF1 (insulin-like growth factor 1). However, as the growth factors are not available in the catalogs, he needed to produce it. He used rat liver cells (buffalo rat liver, BRL) and added the culture medium to the ESCs. After one week, the cells remained undifferentiated in the absence of feeder cells. A. Smith then studied the growth factor that inhibits differentiation. From the BRL medium, he succeeded in purifying this protein. This was not IGF. The N-terminus of the protein was glycosylated, and several sequencing attempts failed. He then attempted to screen the purified protein by means of hematopoietic tests.

A proliferation test was positive and the cytokine was identified, the *LIF* (leukemia inhibitory factor). The LIF owes its name to its ability to induce differentiation of leukemic cells, preventing them from proliferating indefinitely. This factor is expressed in the trophectoderm cells, and its LIF receptor (leukemia inhibitory factor receptor) is located on the cells of the internal mass. As ESCs are derived from the IMC at the blastocyst stage, they are in the presence of the LIF. When cultured *in vitro*, extracting them from this microenvironment amounts to subtracting the LIF source, therefore limiting the proliferation and triggering differentiation.

It was thus demonstrated that the feeder cells produce the LIF, because they produce the soluble form and the form associated with the cell [RAT 90].

A. Smith received the 2010 Louis-Jeantet Prize for Medicine for his fundamental contribution to the understanding of the mechanisms that allow the renewal or differentiation of SCs, an essential step in the development of cell therapy [SMI 10]. Considered one of the world's leading ESC specialists, he explained how these pluripotent cells are formed at the embryonic stage and how they can multiply once cultured. His works paved the way for the development of cell therapy that aims to regenerate damaged organs or tissues. A. Smith is currently Director of the Wellcome Trust Center for Stem Cell Research at the University of Cambridge. In 2000, he received the Pfizer Academic Award for his pioneering works in the field of SCs.

1.3.2. *The ground state of pluripotency*

According to A. Smith, pluripotent SCs can be represented as a ground state or *tabula rasa*. One of his research works consisted of simplifying the medium and the complex environment used for ESC culture and defining the conditions for maintaining pluripotency in order to explore its mechanism. Co-culture can be replaced by LIF, which *activates the STAT3 transcription factor* [NIW 09]. The STAT3 factor belongs to the STAT (signal transducer and activator of transcription) protein family. These proteins play a dual role of signal transmitter and transcriptional activator. In the presence of cytokine, STAT proteins are activated by phosphorylation, and they assemble into dimer, migrate into the nucleus and activate transcription. The specificity of their action depends on the binding of the ligand to its receptor and then on the interaction with other nuclear factors to activate transcription.

Activated in the presence of the LIF, STAT3 acts by reinforcing the circuits that are at the heart of the process, by increasing the expression of the *Krüppel-type factors*, *KLF4* and *KLF5*. If differentiation is committed, activation of STAT3 becomes essential. The commitment of ESCs in the differentiation process is triggered by the *signaling cascade of the MAP kinase* (mitogen-activated protein kinase). These proteins play a role in cell growth and proliferation.

The suppression of the signaling pathway, MAP kinase/ERK (*extracellular signal-regulated kinase*, subfamily of the MAP kinases), maximizes the self-renewal of SCs in the presence of LIF. These observations have made it possible to replace the serum, previously used to prevent this differentiation, with small, highly selective inhibitory molecules. In short, LIF strengthens the transcription stage of pluripotency. *OCT4*, *SOX*, *NANOG* and *KLF* consist of self-regulation factors, which support the pluripotency circuit.

1.3.2.1. *The ground state of embryonic stem cells in mice*

The idea that ESCs are in a *ground state*, identical for all mammalian cells, defines a basal stage of self-replication, allowing cells to survive and proliferate [YIN 08]. This stage is subject to few external *stimuli* and reacts to perturbations of the MAPK (MAP kinase) signaling cascade.

A. Smith demonstrated that inhibition of the MAPK signaling pathway enables maintenance of the pluripotency of ESCs from not only all stems of mice tested but also of rats. Better still, blocking MAPK in the mouse embryo reveals that all cells of the internal mass can become pluripotent, and not just a fraction of them. These findings led to a reinterpretation of the nature of ESCs, rather as cells identical to those of the early epiblast than as an artifact of *in vitro* manipulation. ESCs represent a homogeneous population when the *MAP/ERK signaling pathway is suppressed*. When this pathway is active, the SCs become heterogeneous in terms of gene expression, even if differentiation can be prevented by the LIF. The MAPK/ERK signaling pathway could disrupt the ground state and cause a transition state. In this critical state, the cells are ready to differentiate or return to the ground state.

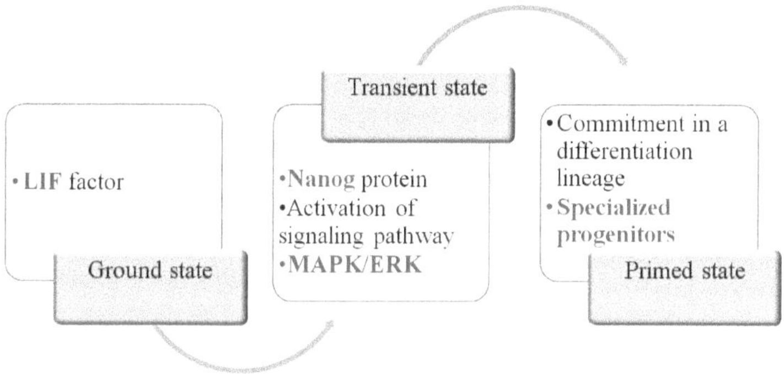

Figure 1.9. *The ground state, or "tabula rasa" according to Austin Smith, followed by an intermediate state of pluripotency before commitment begins*

1.3.2.2. *The ground state of embryonic stem cells in humans*

A. Smith hypothesized that the molecular network, which forms the basis of pluripotency, is conserved in mammals. Nevertheless, unlike those of mice, human ESCs rely on the MAPK signaling pathway to continue to proliferate. This difference is probably due to the fact that currently available human ESCs come from a more advanced stage in development. Given the variable quality of the material available and the lack of tools enabling measurement of the state of maturation of the blastocyst, and in particular that of the epiblast, this research nevertheless remains difficult.

ESCs must first come out of this fundamental state to generate cells capable of committing to differentiation. *The primed state* can be considered as the state following the transition phase, in which several lineages will commit to specific progenitors, hence the idea of successive stages of pluripotency.

1.3.3. *A new distinction between naïve and primed pluripotency in mice*

The pluripotent mouse SCs originate from all of the blastocyst IMCs. They are called ESCs and are the equivalent of preimplantation embryonic cells. Approximately 10 years after the discovery of mouse ESCs, human pluripotent cells were isolated from the blastocyst IMCs. By analogy with mouse ESCs, they were also named ESCs. Nevertheless, they represent both a different morphology and a different pluripotency control mechanism. This suggests that *two distinct ESC origins* exist.

In mammals, embryo preimplantation is broken down into several division and differentiation stages. After segmentation, the zygote is transformed into totipotent daughter cells, called blastomeres. The divisions succeed one another, with the daughter cells more numerous, smaller and forming a sphere, which characterizes the stage of *blastula* (meaning "germ" in Greek), or blastocyst. This sphere is hollow, and the cavity is delimited by a layer of large flattened cells. In this cavity, there is a mass of undifferentiated cells. The first differentiation leads to a spatial distribution and two different futures.

– The *trophectoderm*, from the Greek *tropheus*, meaning "that feeds", is the envelope of the outer epithelial cells at the origin of the placenta. It plays an important role in implantation.

– The *inner cell mass, or inner mass cells* (IMC), called the embryoblast, constitutes the embryonic pole.

The second differentiation divides the IMC into two compartments: the *primitive endoderm*, which will be at the origin of the visceral and parietal endoderm, and the *epiblast*, at the origin of the embryo. When it is implanted in the uterus, the embryo is composed of three morphologically distinct tissues, namely the epiblast, the trophectoderm and the primitive endoderm.

Mouse ESCs are derived from the IMC of blastocysts and represent the *in vitro* epiblast lineage. They are able to remain in undifferentiated state, although the renewal is unlimited. They then differentiate into tissue according to the three embryonic layers *in vitro* and *in vivo*. In addition, the mouse ESCs introduced into a preimplantation embryo are integrated into the embryo and differentiate into tissues, which results in a *chimera*.

In the pluripotent state, mouse ESCs are characterized by the *OCT4* and *NANOG* transcription factors, the expression of *alkaline phosphatase* and specific *surface glycoproteins*, such as stage-specific embryonic antigen-1 (SCA1), and a significant *telomerase* activity. As in the *in vivo* epiblast lineage, female ESCs *have not inactivated the X chromosome*, whereas it will be inactivated during *in vitro* differentiation [ORK 11].

The undifferentiated state of ESCs *in vitro* largely depends on exogenous factors. The first mouse ESCs were cultured in conditioned medium of teratocarcinoma cells or co-cultured with embryonic fibroblasts irradiated in enriched culture medium. It was demonstrated that pluripotency was maintained thanks to the LIF, a cytokine of the interleukin-6 family [NIW 09]. It assures the self-renewal of mouse ESCs by activating STAT3 by the LIF/STAT3 pathway.

1.3.3.1. *Epiblast stem cells*

ESCs are isolated from the IMC of preimplantation blastocysts. In mice, *epiblast SCs (EpiSCs)* are isolated from *post-implantation* epiblast [TES 07]. For ethical reasons, there is *no human-embryo equivalent*. The ESCs from the IMC constitute the reference of the naïve state of pluripotency. It should be kept in mind that the state of pluripotency may depend on the culture conditions and the source of the cells, that is, the state of pre- or post-implantation development of the embryo. However, a functional feature differentiates the two states of pluripotency: *naïve* or *primed*. This is the response of the cells to the inhibition of the *MAPK/ERK signaling pathway*, also called the MEK pathway. Naïve pluripotent SCs tolerate MEK inhibition and remain in *naïve state*, hence the addition of inhibitors of this signaling pathway to maintain the pluripotency of the ESCs. The culture conditions have been described for murine naïve ESCs cultured without serum in the presence of LIF supplemented with two inhibitors, named LIF-2i. On the contrary, to switch to the primed state of EpiSCs, the MAPK/ERK signaling pathway is activated by the presence of FGF2 and activin A [HAN 07a].

Murine ESCs have the molecular characteristics of the IMCs of the embryo and represent naïve pluripotency. A new type of pluripotent cells, derived from the rodent post-implantation epiblast (EpiSCs), presents a primed pluripotency. Molecular and functional differences exist between these two types of pluripotency, which influence the characteristics, functions and safety of these cells.

1.3.3.2. *The conditions of naïve pluripotency*

In order to understand the biology of naïve ESCs and their environment, let us follow the evolution of the culture conditions of these cells. Originally, murine ESCs are cultured in the presence of fetal bovine serum (FBS) and feeder cells, mitotically inactive mouse embryonic fibroblasts (MEFs) [MAR 75]. The LIF is then identified as the key proliferation factor, which activates the JAK-STAT3 signaling pathway, a signal transduction system by the Janus kinase [SMI 88]. The ESCs are then cultured in the presence of FBS and LIF, without MEF.

The conditions for culturing murine ESCs without serum and without feeder cells then use low doses of BMP4 (bone morphogenetic protein 4) and LIF [YIN 03]. They combine the following three inhibitors: an inhibitor of MAPK (PD184352), a tyrosine kinase inhibitor of the FGF receptor (SU5402) and a glycogen synthase kinase-3 inhibitor (CHIR99021). These conditions, called 3i, enable the proliferation of ESCs in the ground state of pluripotency [YIN 08]. However, these conditions induce the inhibition of glycogen synthase kinase 3 (GSK3), which mimics the stimulation of the signaling Wnt pathway by exogenous insulin by activating the PI3K-AKT signaling pathway. New 2i/LIF conditions ("2i" – PD0325901 and CHIR99021) enable improvement of the proliferation of murine ESCs, highlighting the role of autocrine FGF4 (fibroblast growth factor 4), independently of the MEK-ERK signaling pathway [HAN 09].

1.3.3.3. *The conditions of primed pluripotency*

The SCs in primed pluripotency state are EpiSCs, derived from rodent post-implantation epiblasts, maintained in culture in the presence of FGF2 and activin A. These EpiSCs are pluripotent cells capable of differentiating *in vitro* into three embryonic layers. They can generate a teratoma, from the Greek *teratos*, meaning "monster", and the suffix *-ome*, meaning "tumor". Tumor formation is derived from one or more embryonic layers and may be benign or malignant. EpiSCs are *unable to generate chimeras* if they are injected into a preimplantation epiblast. They correspond to a more advanced stage in development. EpiSCs express *OCT4* and *SOX2*, whereas the other pluripotency factors, NANOG, ESRRβ, KLF2 and KLF4, are lower [HAC 14]. EpiSCs present the same epigenetic characteristics as naïve ESCs: they inactivate the female X chromosome and increase DNA methylation by acquiring histone trimethylation (H3K27me3) [GAF 13].

Naïve and primed pluripotent SCs depend on different culture conditions and opposite epigenetic regulators:

– Naïve ESCs are cultured in the presence of serum and LIF (FBS/LIF) or LIF supplemented with two inhibitors of the MEK pathway (2i/LIF). Under these conditions, they tolerate the loss of numerous *epigenetic repressors*, including several methyltransferases (DNA methyltransferase 1,

DNMT1, and methyltransferase-like protein 3, METTL3), which make these cells resistant to differentiation [GEU 15]. In contrast, murine SCs in primed pluripotency state depend on these factors.

– The proliferation conditions of EpiSCs combine the action of two inhibitors. The *GSK3 inhibitor* blocks the activation reaction cascade of β-catenin and stabilizes it in the cytoplasm. Its translocation into the nucleus is inhibited and reinforced by *IWR1, a tankyrase inhibitor*, which stimulates axons 1 and 2, leading to the retention of β-catenin in the cytoplasm. EpiSCs are maintained in proliferation without exogenous supply of FGF2/activin A. As soon as IWR1 is removed from the medium, the transport of β-catenin toward the nucleus increases, which induces the differentiation of EpiSCs [KIM 13]. The mechanism that prevents differentiation thanks to cytoplasmic β-catenin is as yet unclear.

The mouse epiblast SCs, EpiSCs, come from the developmental stage called gastrulation (from the Greek *gaster*, meaning "stomach"). It is characterized by the invagination of one of the poles of the embryo, during this second phase of development. The cells of the embryo forming two cavities are organized into two, then three, separate embryonic layers. The ectoderm is the outer layer, which produces the epidermis and the nervous system. The endoderm is the inner layer, which produces the digestive tract and the accessory glands. The mesoderm forms a real tissue that produces the muscles, the skeleton and the blood vessels. In a gastrula, the embryonic layers are in place, as well as the axes of antero-posterior and dorso-ventral polarity. Organogenesis begins to form the next stage, the neurula.

This process has a *spatio-temporal dynamic*. The differentiation possibilities change according to the moment at which the EpiSCs are sampled. The more the embryonic layers become individualized, the more the differentiation capacities are reduced. The time factor is major, as is the location of the EpiSC. According to their position in the embryo, not only in one of the layers but also within the same layer, the differentiations differ, showing the plasticity of the epiblast. At the moment of gastrulation, the mammalian primordial germ cells (PGC) are determined. The precursors of the PGCs come from the epiblast, more precisely from the proximal epiblast close to the extra-embryonic ectoderm. As of this stage of development, the potential of these progenitors is restricted to the PGC lineage, once their extra-embryonic localization is achieved. As proof, the culturing of a proximal epiblast produces PGCs, whereas a distal epiblast never produces PGCs. By modulating the culture conditions, several differentiations are initiated:

– In the presence of FGF2/activin A, they correspond functionally to the cells of the external embryonic layer, the ectoderm, and more specifically to the primitive cells of the *gastrula* [KOJ 14].

– In the presence of FGF2/IRW1, they correspond to the proximal posterior epiblasts [WU 15].

– The proliferation duration influences the functionality of the cells. EpiSCs maintained for more than 7 days under FGF2/Activin A conditions lose the ability to generate PGCs in response to BPM4 [HAY 11].

More generally, molecular studies into the conversion from the naïve pluripotency state to the primed state have unraveled the key events of reprogramming.

– Under *naïve pluripotency* culture conditions (2i/LIF), EpiCs maintain *hypomethylation levels* similar to those measured in the IMC cells.

– During the transient stage, the level of *methylation of the global DNA increases*, but the regulatory regions (activator and promoter) remain protected from methylation [HAC 13]. The culture conditions correspond to the addition of serum to the medium (FBS/LIF).

– Under conditions of *primed pluripotency*, the culture medium contains FGF2 and activin A, and *the DNA methylation of the regulation elements is functional*.

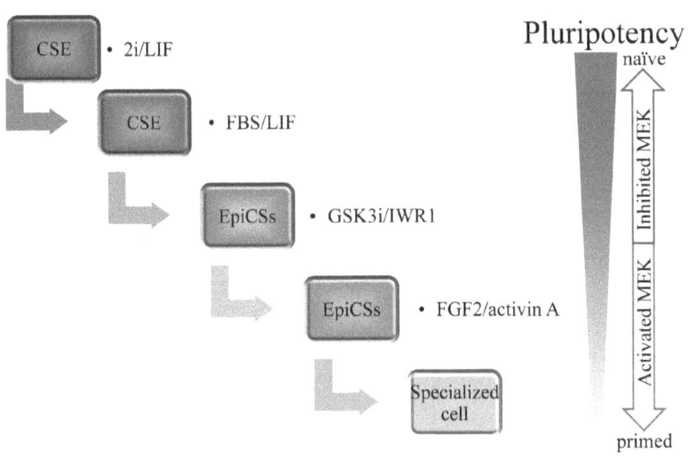

Figure 1.10. *Dynamics of pluripotency states in mice. From naïve pluripotency state to primed state according to the culture conditions. For a color version of this figure, see www.iste.co.uk/arrighi/stemcells.zip*

1.3.4. *What of the pluripotency of human embryonic stem cells?*

Human ESCs isolated from blastocysts present different characteristics and culture conditions from murine ESCs. The LIF does not play a major role, unlike the murine ESCs. On the contrary, FGF2, activin A and TGF-β1 (transforming growth factor-β1) are at the heart of the signaling pathways. They maintain the human ESCs derived from the IMC.

1.3.4.1. *A pluripotency state less primed in human cells*

The differences between conventional human and murine ESCs have been attributed to genetic variation between species. Conventional human ESCs and the pluripotent cells induced by *in vitro* reprogramming share the following characteristics of the primed state [GAF 13]:

– the low level of expression of naïve pluripotency markers, such as KLF17 and DPPA3 (developmental pluripotency-associated protein 3);

– the deposition of the epigenetic mark of trimethylation of histone H3 (Histone H3 trimethyl K27, H3K27me3), inducing the repression of the development genes;

– the loss of the exclusive nuclear localization of transcription factor E3 (transcription factor binding to IGHM enhancer 3, TFE3);

– the loss of pluripotency by inhibition of the MEK-ERK signaling pathway;

– the loss of the overall hypomethylation present in the cells of the IMC;

– the loss of the X inactivation state in the female lines.

Conventional or primed human ESCs are not identical to murine EpiSCs and may be considered as relatively *less primed*. For example, human ESCs do not have an increased expression of FGF5 or N-cadherin (as observed in murine EpiSCs) and strongly express E-cadherin (as observed in murine naïve ESCs) [GAF 13]. Human ESCs strongly express certain naïve pluripotency markers such as NANOG, PRDM14 and REX1 (reduced expression protein), which are not, or are very little, expressed in EpiSCs [CHI 14]. In addition, primed human ESCs are dependent on NANOG and PRDM14, and the absence of these factors induces the differentiation of human ESCs. The distribution of DNA methylation in human ESCs corresponds to that of naïve murine ESCs, cultured under FBS/LIF conditions, rather than those of murine EpiSCs cultured with FGF2/activin A [HAC 13]. In addition, although the localization of TFE3 is exclusively cytoplasmic in murine EpiSCs, it is exclusively nuclear in naïve ESCs cultured in 2i/LIF [BET 13]. In primed human ESCs, the configuration is intermediate and TFE3 is present in the cytoplasm and in the nucleus [GAF 13].

1.3.4.2. *A divergence between human and murine epiblasts*

Although there is no morphological difference between human and murine blastocyst, molecular differences exist at the cell level. The human cells of the epiblast IMC do not express the KLF2 and ESRRB genes, which are important pluripotency factors in mice. KLF17 would be human-specific in the IMC. Non-human primate, marmoset, pluripotent epiblasts are very close to human epiblasts, but remain very different from mouse epiblasts. Marmoset and man have IMC signatures in common: the loss of the transcription factor KLF2, nuclear receptor NR0B1 (subfamily 0, group B, member 1) and BPM4, with high levels of expression of NODAL [BOR 15].

At the post-implantation stage, major differences exist between human and rodent embryos. Those of rodents have a cylindrical shape, whereas in humans, the epiblast resembles a flat disk, similar in shape in the majority of mammals. As early human epiblast analysis is not possible, the study of non-human primates can provide new elements.

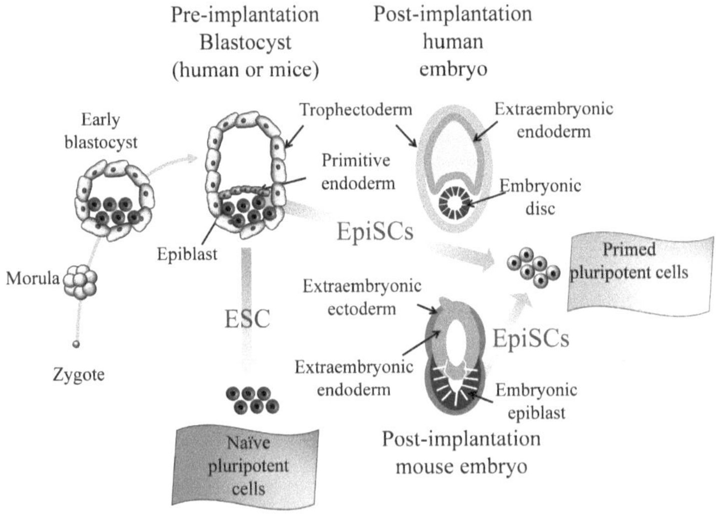

Figure 1.11. *Pluripotent, naïve and primed (EpiSC) SC line in humans and mice. For a color version of this figure, see www.iste.co.uk/arrighi/stemcells.zip*

These SCs of the epiblast have one more differentiation step than the SCs of the preimplantation blastocyst. This demonstrates the existence of a pluripotency control mechanism. This fine regulation conditions the sequence of the stages of the embryo's future development. There therefore exist microenvironment *stimuli* that will set the tempo of the specialization. The first stage of multiplication of identical SCs results from a dynamic equilibrium between the transcription factors maintaining "hyper naïve" pluripotency and those that push toward its exit, driving the beginning of the lineage of differentiation. *In vivo*, the duration of this phase is ephemeral. *In vitro*, it can be influenced one way or the other and can induce the pluripotency of the already-primed cells.

1.4. Induced pluripotent stem cells

The winner of the 2012 Nobel Prize in Medicine together with John B. Gurdon, Professor Shinya Yamanaka obtained for the first time, in 2006, *induced pluripotent stem cells (iPS)*. Starting from a differentiated skin cell, Yamanaka induced an undifferentiated cell, or SC, similar to that found at the embryonic stage. Yamanaka's early works focused on the Fbxo15 gene (F-box only protein 15), which he identified as a specific gene of mouse ESCs. He did not, however, observe any particular phenotype on mice, nor on Fbxo15-deficient ESCs, but retained this line, which was to prove very useful in the generation of iPSs.

Yamanaka currently heads the Center for iPS Cell Research and Application at Kyoto University.

1.4.1. *The discovery of induced pluripotent stem cells*

While the scientific community is questioning the mechanisms of differentiation of ESCs into somatic CSs, Yamanaka sees two major difficulties in the use of ESCs, one ethical and the other immunological.

The first element is that these ESCs originating from the human embryo raise the question of the embryo's status. They crystallize the *ethical issues* of research. The status of the embryo and embryo cells raises ethical considerations that must be integrated into research concerning human embryonic stem cells and *in vitro* human embryos. These ESCs are rare and valuable for two reasons. First, they are few in number in an embryo:

approximately 30 totipotent cells. Second, the number of embryos is limited because they come from an interrupted pregnancy. In addition to the ethical limit, the source of ESCs is a second limiting factor. This bioethical consideration and the texts that govern research into ESCs will be developed in Chapter 5.

The second element is that such ESCs transplanted into humans could induce a transplant rejection. The *immunological* feasibility poses a problem. The body's immune response against a foreign body induces the rejection of the transplanted tissue. To avoid rejection of the transplant, the donor and recipient must be compatible, or even better, be the same person, which is impossible with human ESCs.

With the aim of generating pluripotent cells close to ESCs *without using an embryo*, Yamanaka has begun focusing on nuclear reprogramming. This technique was demonstrated by Sir John Gurdon in 1962, who generated frogs from tadpole intestinal cell nuclei transferred into enucleated eggs [GUR 62]. Then, in 1997, Sir Ian Wilmut's team unveiled Dolly the Sheep, the first mammal to be cloned by nuclear material transfer [WIL 97]. Yamanaka's hypothesis is based on the existence of *factors maintaining the pluripotency* of mouse ESCs, which could induce pluripotency in somatic cells.

1.4.1.1. *The first induced pluripotent cells in mice*

The most important step in the generation of iPSs is the identification of pluripotency factors. The list of candidate genes established *in silico* compares the ESC cDNAs to those of the somatic cells. By subtraction, the cDNAs coding for ESC genes are designated as ES cell-associated transcripts (ECATs) [MIT 03].

The strategy used, gene trapping, enables the *loci* whose transcription is activated during a given simulation to be revealed. The selection cassette used is under the control of Fbxo15 promoter (F-box only protein 15), only active in ESC. This promoter is associated with two revealing genes (Fbxo15-β-geo). The β-*gal* gene codes for β-galactosidase and the *geo* code gene for a fusion product between β-galactosidase activity and neomycin-phosphotransferase conferring resistance to geneticin (G418).

The next step is to introduce these genes, identified as *possible reprogramming factors*, a total of 24 candidate genes, into mouse embryonic fibroblasts (MEFs) and to verify whether they actually induce pluripotency. The retroviral transduction system is used for its high efficiency. The candidate genes are introduced, one by one, into the selection cassette under the control of the Fbxo15 promoter, as well as all 24 genes. None of the factors integrated in isolation enable the cells to survive in the presence of an antibiotic. In contrast, 22 colonies survive the antibiotic in the construct corresponding to all 24 genes. These colonies resemble those of ESCs and do not resemble those of the MEFs from which they originate. Then, Yamanaka's team decided to reduce this list of 24 candidate genes by removing just one of the factors and testing all remaining 23 factors, thinking that by generating ESCs from the 23 candidates, the one that is removed is not crucial. This strategy enables the list to be reduced to 10 genes; it is the first synthesis of iPS cells (iPS-MEF10). The second test, the elimination of four genes coding for *OCT3/4*, *SOX2*, *KLF4* and *MYC* inhibits the formation of colonies. Finally, the ability of these four genes to induce pluripotency is confirmed by the morphological and functional reprogramming criteria:

– the *morphology* and gene expression of pluripotency markers;

– *unlimited growth*;

– induction of *tumors* (teratocarcinomas) by injection of the cells; transplanted into immunodeficient mice (nudes), these cells form teratomas containing various tissues from the three embryonic layers;

– injection into blastocysts gives *chimeric* mice.

Of the numerous combinations, the following *four factors are essential*: *OCT3/4*, *SOX2*, *KLF4* and *c-MYC*, also called *OSKM factors*.

In 2005, Yamanaka generated ESC-type cells with these four factors, naming them *induced pluripotent stem cells*, or *iPSs*. After confirmation, the results were published in 2006 in the journal *Cell* [TAK 06]. This discovery surprised the majority of researchers, because just four factors were enough in order to reprogram somatic cells to the embryonic state. Other laboratories at the Massachusetts Institute of Technology and Harvard University then produced mouse iPS using the same protocol.

1.4.1.2. *The first human-induced pluripotent cells*

After generating mouse iPSs, Yamanaka's team attempted to reprogram human somatic cells from the four transcription factors via viral vectors, succeeding in 2007 [TAK 07]. They demonstrated that the OSKM factors were sufficient to generate human iPSs. At the same time, James Thomson's team proposed four different factors [YU 07], two of which are common, OCT3/4 and SOX2, whereas the other two, NANOG and LIN28, are different. The two reprogramming protocols are commonly used.

1.4.2. *The key factors of pluripotency*

The following four proteins are transcription factors: OCT3/4, SOX2, KLF4 and c-MYC. They are at the origin of a sequence of epigenetic events that modify the organization of chromatin and DNA methylation, transforming the differentiated cell into a pluripotent cell. The cell lines thus obtained present properties that are very similar if not identical to human ESCs and are capable of differentiating *in vitro* into any cell type.

– *OCT3/4* (octamer-binding protein 3/4) is a finely regulated transcription factor, associated with a large number of pluripotency-specific genes, themselves closely related to SOX2. OCT3/4 is the key factor in SC self-renewal. The octamer of this family of transcription factors is the 8-nucleotide sequence of DNA (ATTTGCAT).

– *SOX2* (sex-determining region Y-box 2 or SRY box-containing factor 2) is necessary for embryonic development and to prevent ESC differentiation. Several genes associated with pluripotency are co-regulated by SOX2 and OCT3/4. Nevertheless, SOX2 can also cooperate with other transcription factors, for example with NANOG, to activate the transcription of the pluripotency markers.

– *c-MYC* (cellular myelocytomatosis oncogene) is a proto-oncogene gene overexpressed in certain human helix-loop-helix cancers, involved in various cellular functions. It is the STAT3 effector in the LIF signaling pathway. c-MYC is a substrate of GSK3b in the Wnt signaling pathway. In iPS cells, c-MYC can compensate for the anti-proliferative effect of KLF4.

– *KLF4* (Krüppel-like factor 4) can not only act as an oncogene, but also as a tumor suppressor. Similar to c-MYC, KLF4 is a STAT3 effector in the LIF pathway, and its overexpression inhibits ESC differentiation. KLF4 regulates the Lefty1 transcription factor with OCT3/4, but the role of OCT3/4 cofactor is limited to certain target genes. KLF4 can suppress p53, a NANOG inhibitor.

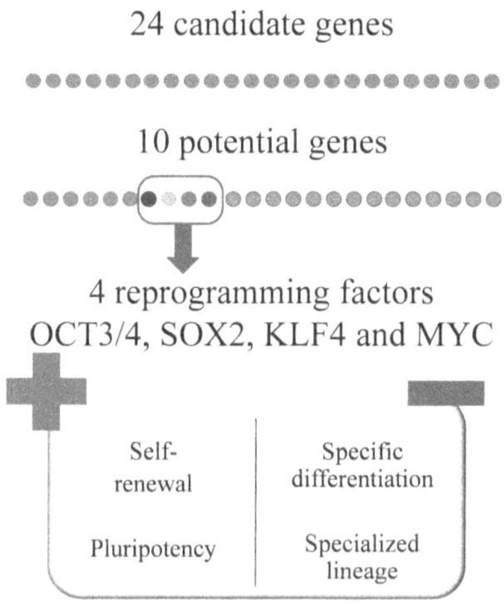

Figure 1.12. *Yamanaka's strategy in the discovery of reprogramming factors and iPS generation. For a color version of this figure, see www.iste.co.uk/arrighi/stemcells.zip*

Recent works showed that differentiated cells can be reprogrammed into embryonic-type pluripotent cells, thus providing a source of pluripotent SCs specific to each patient, hence the benefit of understanding the molecular basics of the regulation of the fate of SCs and their therapeutic applications. However, before these SC-based therapies can be applied routinely and in complete safety, the role of the transcription factors – OCT4 (or POU5F1), NANOG and SOX2 – in maintaining SC pluripotency must be explored.

The maintenance of ESC pluripotency is controlled by a set of signals from the extracellular environment and the dynamic equilibrium of cellular transcription factors. Despite the different signaling pathways between humans and mice, the key transcriptional circuit is remarkably preserved. The transcription factors OCT4, NANOG and SOX2 form a central module maintaining the identity of ESCs in humans and mice. This module is rich in positive controls. In particular, OCT4 and SOX2 form a heterodimer, which positively regulates the expression of POU5F1 (which codes for OCT4), SOX2 and NANOG. In addition, NANOG interacts directly with OCT4 and positively regulates the expression of the three genes. Thus, these three transcription factors regulate their own expression and coordinate that of the other two by protein–protein or protein–DNA interaction loops.

These three genes are also involved in the development and repress the expression of genes involved in the commitment of cells in a lineage, as for example in the development of the following embryonic layers:

– trophectoderm: HAND1 (heart and neural crest derivatives-expressed protein 1) and eomesodermin (EOMES);

– ectoderm: LHX5 (LIM homeobox 5), OTX1 (orthodenticle homolog 1) and HOXB1 (homeobox protein-b1);

– mesoderm: MYF5 (myogenic factor 5), T (brachyury protein homolog), GSC (goosecoid);

– endoderm: FOXA2 (forkhead box A2) and GATA6 (GATA-binding protein 6).

In parallel, OCT4, NANOG and SOX2 activate genes associated with *self-renewal* and *pluripotency*, including ESC transcription factors such as TCL (T cell leukemia/lymphoma), TBX3, REST, ZIC3, HESX1 (expressed in ESCs), STAT3, REX1 (or ZPF42), SALL4, TCF3 and DAX1 (or NROB1). OCT4, NANOG and SOX2 play a central role in the maintenance of the identity of ESCs. The expression of this trio maintains the cell in a state of self-renewal by the activation of other ESC-specific genes and the inhibition of genes associated with the commitment of the cells in a lineage. Loss of expression leads to loss of ES phenotype (self-renewal) and commitment to differentiation.

Cell reprogramming can be represented according to *the Waddington epigenetic landscape diagram* [WAD 57]. This representation illustrates the development of an organism by the rolling of the ball in one of the valleys, which corresponds to the path of the development of an organism, culminating in an end. Biological finality is expressed in this way in a mathematical context. Stable regimes likely to be reached by a system are called attractors. Several pathway changes can appear in a development path. The activation or non-activation of a gene can modify the path chosen. Cell reprogramming comes down to guiding the cell in this landscape from one position to another, where several attractors coexist (minimal position) and each position has its own molecular signature.

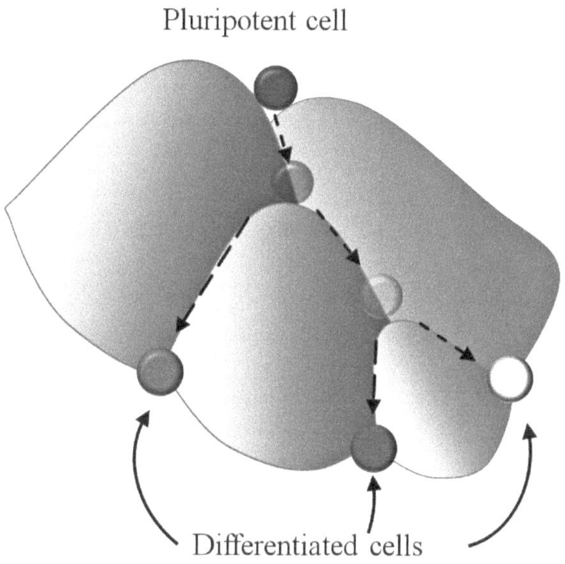

Figure 1.13. *According to Waddington's epigenetic landscape diagram, the development of an organism corresponds to the path of a ball from the top, representing the pluripotent state, to the bottom, differentiation [WAD 57]. For a color version of this figure, see www.iste.co.uk/arrighi/stemcells.zip*

1.4.3. *The reprogramming mechanism*

Reprogramming occurs in two phases, one early and the other late. During the early phase, important events are the suppression of somatic cell genes, the transition from the mesenchymal to the epithelial cell and the metabolic shift from oxidative phosphorylation to glycolysis. At the same time, damage to the DNA, oncogene-induced senescence and apoptosis block the progression of reprogramming. During the late stages of reprogramming, the genes associated with pluripotency are activated. Developmental genes are repressed as well as those of the specific transcription factors of certain tissues, some promoters of which are trimethylated by histone H3 (Histone 3 Lysine 4, H3K4 and Histone 3 Lys27, H3K27).

The *OSKM transcription factors* function as “pioneer factors”, binding to regions of the chromatin, becoming inaccessible to other factors, leading to activation or inhibition of gene expression [SOU 12].

During the early reprogramming phase, OSKM factors bind to the stimulatory and promoter genes of the somatic genes and to the pluripotency-associated genes (PAGs), but cannot bind to late PAGs. MYC, for example, binds to genomic *loci* methylated by H3K4, a chromatin marker in open state. These *loci* are stimulatory or promoter genes that determine the identity of the somatic cell and reduce the somatic genes to silence [SOU 12]. OSKM factors bind simultaneously to the stimulatory and promoter genes of pluripotency, inducing their expression. The cells partially reprogrammed during the early phase then enter into the second phase. OSKM can be replaced by the effectors LIN28, SALL4, ESRRβ and NANOG, in mouse iPSs, which would explain the low reprogramming efficiency [BUG 12].

Later, the OSKMs start the expression of late genes. In iPSs, exogenous OSKM factors are no longer expressed and early and late PAGs are fully activated. Stoichiometry between factors is important during reprogramming. In the early phase, the levels of OCT3/4 and SOX2 are high and low, respectively.

The fully reprogrammed iPS cells present the same characteristics as the ESCs. In contrast, cells that continue to express exogenous reprogramming factors show the characteristics that are lower than those of the ESCs (similar morphology, rapid proliferation, activation of pluripotency-associated genes, teratoma formation and chimeric contribution) [TAK 06]. In addition, these *partially reprogrammed cells* comprise epigenetic traces of somatic cells and may present a defective differentiation [TAK 06]. This represents the major obstacle to the practical application of iPSs. Some approaches have enabled an increase in *reprogramming efficiency*, for example, by increasing the reprogramming duration. Another improvement is to introduce reporter systems dependent on the Oct*3/4* or Nanog promoter, rather than the Fbxo15 selection cassette. As the expression of OCT3/4 and NANOG appears after that of Fbxo15, this method can be used to select more mature (mouse) iPSs. For human iPSs, the reprogramming efficiency can be increased by the co-expression of reprogramming stimulators. The culture medium can be optimized. A treatment with inhibitors of MEK and glycogen synthase kinase 3, medium 2i, stimulates pluripotency (homogeneous population in ESC) and increases the reprogramming efficiency of mouse iPSs [YIN 08].

The major problem encountered with the first generation of iPSs is the use of retroviruses. These *vectors are integrated into the genome* of the host cells and cause the cleavage or aberrant activation of neighboring genes, with the risk of reactivating the reprogramming factors themselves. For example, the reactivation of MYC, one of the OSKM factors, sporadically induces tumor formation in chimeric mice [OKI 07]. The solution to this problem is based on methods that do not integrate the genome, such as the Sendai virus [NIS 11] and synthetic mRNAs [WAR 10]. The protocols quickly remedied the disadvantages detected during the first experiments; it is a rapidly evolving technique. In parallel, translational research tested the concept of this induced pluripotency for therapeutic purposes in order to demonstrate its feasibility.

1.4.4. *Proof of concept of a possible therapeutic utility to induced pluripotent cells*

1.4.4.1. *Model of mouse with sickle-cell anemia*

Jacob Hannah and his team at the *Whitehead Institute for Biomedical Research* in Cambridge, Massachusetts, used iPS cells to treat mice with *sickle-cell anemia*, also known as drepanocytosis, from the Greek *drepanon*, meaning “sickle” [HAN 07b]. This genetic disease of hemoglobin is due to the mutation of the gene encoding β globin. The hemoglobin is referred to as hemoglobin S (hβS), standing for “sickle”. When the oxygen concentration decreases in the blood, the hemoglobin polymerizes and gives the red blood cell a sickle shape.

iPS clones are obtained from mouse fibroblasts with sickle-cell anemia after the expression of four cDNAs coding for OCT3/4, SOX2, KLF4 and c-MYC (with a lox sequence permitting excision of the transgene), inserted into the genome via a murine oncoretrovirus (murine leukemia virus, MLV). These iPS clones, whose karyotype is normal, have all the characteristics of the pluripotent cells described and generate viable chimeras after their transfer into a re-implanted blastocyst in the uterus of a pregnant female. One of the clones is treated with an adenovirus carrying *cre* to excise the exogenous oncogene, *c-myc-lox*, then the cells are corrected by a homologous recombination of the mutated hemoglobin and the normal adult hemoglobin (hβA). Differentiation is induced in the embryonic bodies. The commitment to the hematopoietic lineage is stimulated by the transduction of the HoxB4 cDNA via a murine oncoretrovirus, MSCV (murine stem cell virus). These hematopoietic SCs derived from the corrected iPSs (hβS/hβA) are then amplified on the OP-9 stromal line and then injected into irradiated mice with hβS/hβS falciform anemia. The hematopoietic system of these mice is reconstituted (70%) and, in particular, *the red blood cells express normal adult hemoglobin*, and the anemia is corrected.

1.4.4.2. *Okano's works on bone-marrow lesions*

Dr. Hideyuki Okano's team at the Keio University Medical School, Tokyo, succeeded in repairing *damaged spinal cords in mice* using iPSs. These iPS cells are created from human fibroblasts through the transduction of four genes (OCT3/4, SOX2, KLF4 and c-MYC) [NOR 11]. By the application of a specific culture medium, these cells were reprogrammed into precursors of cells of the nervous system. Once these precursors were grafted within the spinal cord of mice rendered disabled, they migrated and differentiated into *neurons*, *astrocytes* and *oligodendrocytes*. Several positive effects were then observed: formation of synapses between the neurons induced by the iPS cells and mouse neurons; expression of neurotrophic factors; marked angiogenesis; axonal growth and myelin production in the injured areas. Approximately 4 months after this treatment, all 18 mice, disabled and unable to move around before the experiment, were able to regain satisfactory mobility and walk at a speed of 8 cm/s. As a result of these works, the repair process continued and none of the subjects developed cancer [NOR 11].

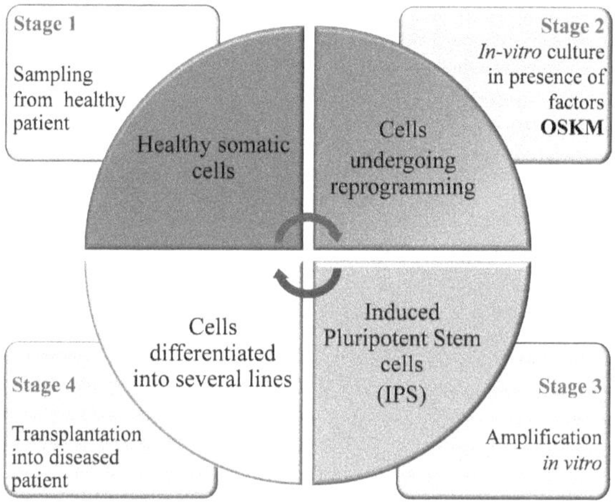

Figure 1.14. *The principle of cell therapy using iPS cells, according to Yamanaka. iPS preparation involves quality control followed by the freezing of clinical-grade iPS stocks according to good manufacturing practice (GMP). For a color version of this figure, see www.iste.co.uk/arrighi/stemcells.zip*

2

Stem Cells as a Necessary Experimental Platform in Medical Research

> *"Researchers need the donation of our samples to find new therapies."*
>
> (Thierry Lhermitte, Actor, Sponsor of the French Foundation for Medical Research (FRM), 2015)

Stem cells (SCs) are present in the embryo from the early stages of fertilization and are at the origin of all of the tissues of the organism. *In vivo*, they will enable the development of the zygote egg via the embryo, then the fetus and finally the whole organism; the living being. This final organism possesses all types of adult SCs, which will assure the replacement of damaged tissues in the event of injury or illness throughout life.

Extracted from the organisms, these SCs can be cultured *in vitro*, where they conserve their fundamental properties, that is to say, the self-renewal and differentiation capacities. They then offer the study model of a given tissue or of an organ of the living organism. In a microenvironment favorable to proliferation, and according to the stage studied, they reproduce in a culture dish *in vitro*, the process which takes place *in vivo*. The major applications of SCs are associated with the development of diseases and avenues for treating them.

First, it is the *modeling of pathologies in vitro*, a veritable technological tool that enables us to understand the disease process.

Second, SCs serve as a *pharmaceutical screening* tool. As the targets of the disease and its mechanism are identified, new treatments can be designed. New chemical molecules are synthesized, whose active ingredient will be directly tested in order to prove its effectiveness on cells from affected SCs. These medications need to be effective and above all non-toxic. In the early phases of clinical trials, high-risk molecules are discarded.

Third, the use of SCs in *predictive toxicology* makes it possible to characterize the toxicity of candidate drugs in humans and to validate the preclinical stages of pharmaceutical research.

2.1. A tool for modeling human pathologies

Modeling a pathology also involves knowing its cause and consequences in order to invent its treatment. Several of these steps have been achieved thanks to SCs.

Knowledge of a pathology first of all requires identification of the patient's symptoms and the drawing up of a complete clinical picture with the vital constants and biochemical data. Interpreting these crossed data directs the clinician to the diagnosis. At cell level, this amounts to defining phenotypes on a physiological, molecular or genetic scale. Based on this targeted characterization of a pathology, the SCs will bring a modeling tool that will enable a strategy to fight the disease to be designed. The strategy will be different according to whether it is a mutation on a gene (monogenic disease) or on a set of genes (complex disease). Several modelings have been invested thanks to SCs in the study of neurological diseases, such as Alzheimer's disease, as well as those which develop late. The key problem is then to identify age measurement markers *in vitro*.

For *pathology modeling in humans*, patient-specific iPS cells are obtained from adult skin SCs. These iPS cells are genetically modified to return to immature state (reprogramming), similar to ESCs. Researchers can then re-differentiate them into various specialized cells (heart, liver, retinal or nerve cells, etc.) that will be used as a study model. They then have an *unlimited source of diseased cells* to study the mechanisms of the pathology, to try to correct the mutation, to test therapeutic molecules or to evaluate the toxicity of a product.

There are already iPS cell lines available to study many diseases such as amyotrophic lateral sclerosis, Parkinson's disease, type-1 diabetes mellitus, Huntington's disease, Down's syndrome, severe combined immunodeficiency, Lesch–Nyhan syndrome, Gaucher's disease, Shwachman–Bodian–Diamond syndrome and Duchenne and Becker muscular dystrophies.

2.1.1. *Identification of the phenotype of the pathology thanks to stem cells*

The modeling of human diseases *in vitro* requires *the identification of the specific phenotype* of each disease. These phenotypes are of three levels: molecular, cellular and physiological. Certain models require complex culture techniques recreating the interactions between the different cell types in a three-dimensional microenvironment, called *3D culture*.

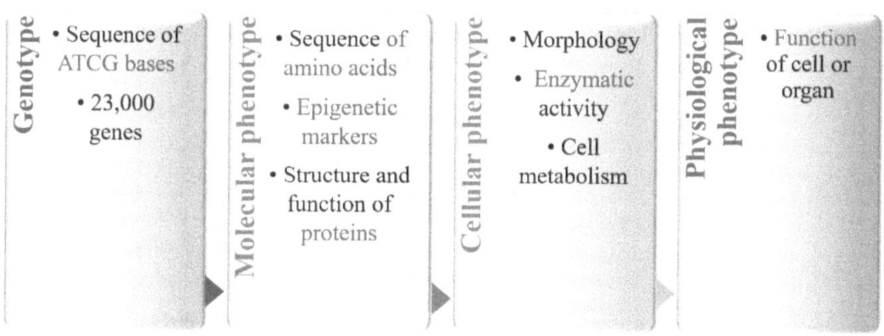

Figure 2.1. In vitro *human disease modeling requires characterization of the specific phenotype of each disease*

The *molecular phenotype* analyzes the expression of the molecules produced in the pathological cell, thanks to transcribed RNAs and proteins. Epigenetic markers also exist, that is to say, transformations that operate during the transcription steps and modulate the final expression of proteins.

For example, methylation of DNA results in histone modification, which limits the overall gene expression or that of the proteome. Molecular phenotypes are widely used in high-throughput screening because they can be measured automatically. In 2012, the restoration of the expression of the IKK (complex-associated protein or IKBKAP) protein, whose gene is mutated into familial dysautonomia (FD), was automatically tested on 6,000 compounds by reverse quantitative transcriptase in real time (qRT-PCR), in order to study treatment avenues [LEE 12].

The *cellular phenotype*, often explored *in vitro*, includes the study of morphology, protein aggregation or apoptosis. Morphology changes are often seen in differentiated neurons from iPSs of patients with neurological disorders. As another example, patients with glycogen storage disease type-1a (GSD1a) cannot maintain glucose homeostasis and present intracellular accumulations of glycogen and lipids. Differentiated hepatocytes from patients with GSD1a present phenotypes, evidenced by periodic acid or Schiff staining.

The *physiological phenotypes*, the most complicated to evaluate, are recorded in neurological and cardiovascular diseases. They consist of measuring the contractibility, the activity of an ion channel, and use multi-electrode networks or patch-clamp analyses of neurons and cardiac muscle. The cardiomyocytes differentiated from iPSs of patients with congenital long QT syndrome show prolonged action potentials and greater sensitivity to isoprenaline (isoproterenol), which is a β-adrenoreceptor agonist. Similarly, neurons derived from patients with amyotrophic lateral sclerosis (ALS) show high hyperexcitability and discharge frequency compared to control neurons.

Certain models of diseases require complex culture techniques recreating the *interactions between the different cell types* in a three-dimensional microenvironment, called *3D culture*. The co-culture of neurons and glial cells was necessary in order to model the ALS. Glial cells that mutated onto the SOD1 gene have been shown to be toxic to spinal motor neurons [DIG 07]. This toxic effect of the glial cells had not been demonstrated when the neurons were cultured in the presence of SOD1-mutated fibroblasts.

It is important to note that the differentiation of SCs in a 3D culture medium can reveal phenotypic characteristics that can be used to model the disease. For example, *brain organoids* developed from the iPSs of patients with microencephaly exhibit not only a premature neural differentiation but also a smaller size than control organoids. In addition, they present an aberrant orientation of the radial glial cells. These structure phenotypes are only detected in 3D culture.

2.1.2. *A multiplicity of possible approaches to pathologies*

Affected SCs used in pathology modeling can have two origins, as they are sampled from:

– the embryo that carries the disease. During preimplantation diagnosis, the ESCs are isolated and preserved in a biobank;

– the patient's somatic cells that are then reprogrammed. They generate *patient- and pathology-specific iPSs.* These SCs will be cultured, amplified and differentiated. The molecular and cellular phenotypes will then be explored. Several types of genetic disorders exist that characterize diseases; these first criteria are to be taken into account during modeling, depending on whether a single gene or multiple genes are involved. These diseases are said to be:

– *Monogenic* because a single gene is altered; this is the case of Lesch–Nyhan syndrome and fragile X syndrome, modeled thanks to human SCs by mutagenesis of the gene associated with the disease. Affected ESCs have been isolated from blastocysts after preimplantation diagnosis at 10 weeks or derived from somatic cells sampled from the patient and then reprogrammed to iPSs.

– *Chromosomal* through the loss or addition of chromosome or chromosomal region (Down's syndrome and Turner's syndrome). They have been modeled using ESCs bearing chromosomal aberrations or isolated from embryos after preimplantation diagnosis at 12 weeks or through the reprogramming of the patient's somatic cells, containing these chromosomal disorders.

– *Complex*, involving several genes such as schizophrenia and autism (autism spectrum disorder, ASD). Complex diseases that cannot be diagnosed early or reproduced by genome editing techniques are modeled by reprogramming patient iPS cells. The difficulty in diagnosing complex diseases lies in their genetic correction. Although genome editing techniques are very effective, they remain limited to a small number of genes. Thus, complex diseases that involve several gene mutations cannot be corrected in the same way as monogenic diseases. However, while the same disorder can come from a single gene or from several genes, researchers very often choose the monogenic model for practical reasons. This is the case for amyotrophic lateral sclerosis (ALS) with mutations of superoxide dismutase 1 (SOD1) or models of Parkinson's disease, whose mutation is known (parkin RBR E3 ubiquitin protein ligase, PARK2).

2.1.2.1. *The complex case of Alzheimer's disease*

Alzheimer's disease is the most common degenerative disease in the elderly. The cause of this disease is not known; it results from a combination of numerous genetic and environmental risk factors. Several studies have generated human-induced pluripotent SC models of Alzheimer's disease from fibroblasts. These models have reproduced the phenotype of the disease by increasing the amyloid peptides β (Aβ), phosphorylation of Tau protein and the presence of large endosomes.

Based on neurons from iPS cell lines, studies have focused on the *SORL1 gene*. The sortilin-related receptor 1 (SORL1) gene encodes the apoliprotein neuronal receptor, which regulates the transformation of amyloid precursor protein (APP) to amyloid-β (Aβ); the main component of the amyloid plaques present in patients' brains [YOU 15].

Cell phenotypes of genomic variants are difficult to detect in post-mortem tissues because several cell types are present in a diseased brain. Young's hypothesis was to purify a culture of human neurons containing the endogenous genome of an individual, in order to detect differences in phenotypes. Fibroblasts of patients with Alzheimer's disease were sampled by biopsy. These fibroblasts were subjected to the four reprogramming factors and led to lines *of induced pluripotent SC models of Alzheimer's disease*. They were differentiated into neural SC lines and purified into neurons by *fluorocytometry* (fluorescence-activated cell sorting, FACS). All of these iPSs are euploid and can differentiate according to the three embryonic layers, namely endoderm, mesoderm and ectoderm. The differentiated neurons are a mixture of glutamatergic, GABAergic and cholinergic subtypes and are electrophysiologically active. In order to know whether the genomic variants of SORL1 influence the level of basal expressions of neurons derived from patients' iPS, the fibroblasts were genotyped and associated with the presence or absence (protection) of a risk factor for the disease.

The absence of a SORL1 expression has been reported in Alzheimer's disease. Recently, neurons from iPS cells of patients with Alzheimer's disease have made it possible to study the role of SORL1 variants [YOU 15]. In neurons differentiated from healthy individuals and patients with Alzheimer's disease, BDNF (brain-derived neurotrophic factor) induces the expression of SORL1, which decreases Aβ. This response is correlated with a 5'-end variant gene of SORL1, which therefore determines a protective haplotype. These observations show the *complexity of Alzheimer's disease, because the same variants are present in both healthy and affected people.* The expression variation of SORL1 induced by BDNF is regulated by common genetic variants and can explain how the genetic variation in this locus contributes to the individual risk of developing Alzheimer's disease.

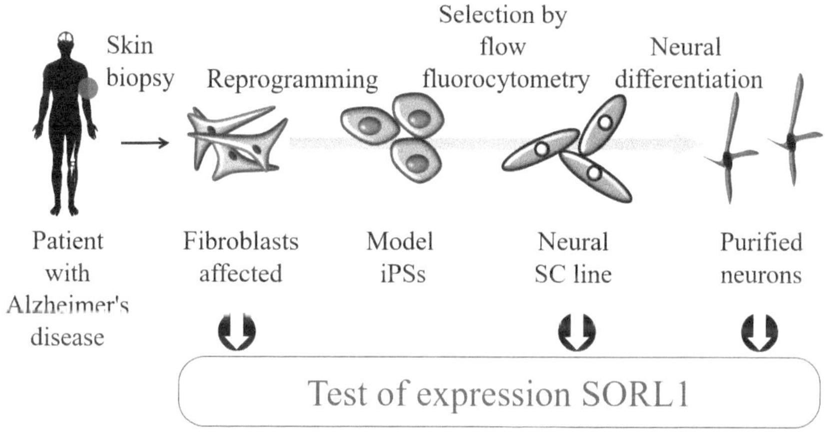

Figure 2.2. *Experimental design and cell lines in the study of molecular phenotypes caused by the risk factor SORL1 in Alzheimer's disease using human iPSs [YOU 15]*

2.1.2.2. *The case of late-onset diseases*

The occurrence of late-onset neurological diseases such as Parkinson's disease, the most common after Alzheimer's disease, is increasing because of increasing life expectancy. To model neurological pathologies, iPS technology uses skin cells from patients, reprogrammed at the initial pluripotent stage and then differentiated. Pluripotent SCs must be able to mimic cellular aging *in vitro* if the disorders appear late, hence the benefit of studying the age-related process with progeria syndrome and in parallel the mechanism of Parkinson's disease.

2.1.2.2.1. Modeling Parkinson's disease

Parkinson's disease is characterized by the progressive loss of dopaminergic neurons. Although this disease is based on a complex genetic basis, certain patients who have a triple mutation of the α-synuclein gene (SNCA) develop symptoms early. iPSs from these patients allowed researchers to identify *in vitro* specific phenotypes after the *differentiation of iPSs into dopaminergic neurons*.

Parkinson's disease affects the motor system and is also manifested by non-motor clinical parameters. The loss of neurons is accompanied by intracellular protein inclusions called "Lewy bodies" or "Lewy neurites", which contain α-synuclein (αSyn). The nonsense mutations and the high copy number of the SNCA gene encoding αSyn (PARK1/4 locus) cause the early onset of Parkinson's disease.

The comparison between lines of iPSs derived from patients with this disease and healthy individuals is possible, but unfortunately, phenotypic differences not specific to the disease can appear. This is due to the high variability between iPS clones, which increases when the cell lines are genetically different, hence the importance of manipulating mutations on a single gene during pathological modeling.

Oliveira's studies have shown the impact of *αSyn overexpression on the differentiation capabilities and phenotypic signature* of two iPS clones derived from the same patient with triplication of the SNCA gene [OLI 15]. In parallel, these results were compared to two control lines from healthy individuals and a line from neuroprogenitors of the patient infected by a lentivirus incorporating an shRNA targeting the SNCA gene, in order to inactivate it.

The SNCA gene triplication lines have a reduced ability to differentiate into dopaminergic or GABAergic neurons, reduced neurites and a low neuronal activity compared to the control cultures. This delayed-maturation phenotype was confirmed by gene expression profiles, revealing a significant decrease in transcribed RNAs of neuronal differentiation genes such as delta-like homolog 1 (*DLK1*), gamma-aminobutyric acid type B receptor subunit 2 (GABABR2), nuclear receptor-related protein 1 (NURR1), G-protein-regulated inward-rectifier potassium channel 2 (GIRK-2) and tyrosine hydroxylase (TH).

The overexpression of αSyn is doubled in case of triplication of the gene and will interfere with the differentiation of the neuronal progenitor cells, involved in adult neurogenesis, which explains the progression of Parkinson's disease.

2.1.2.2.2. Modeling of progeria syndrome

The key problem in modeling late-onset diseases lies in identifying markers of age measurement *in vitro*. Cell markers related to progeria syndrome can provide a solution to this problem. Several studies have shown that the criteria associated with age are lost during reprogramming to iPS. The telomeres lengthen, the mitochondria are active and the senescence markers disappear in the iPSs derived from older donors, suggesting rejuvenation of the cells during reprogramming.

An alternative approach to aging is to *target age-related markers*. Hutchinson–Gilford *progeria syndrome* is a genetic disorder characterized by early aging of all of an individual's tissues, leading to death before reaching adulthood. Mutations in the LMNA gene, which encodes the lamin A of the nuclear envelope, activate a cryptic splice site that produces a truncated transcript called *progerin*. This progerin, accumulated in the nucleus, prevents the ripening of lamin A, which interferes with the multiple processes of chromatin organization in the nucleus, such as heterochromatin formation, DNA repair, cell cycle, gene transcription and maintenance of the telomeres. Small amounts of progerin are expressed in healthy individuals as well as in the fibroblasts of donors of a similar age.

Miller's works related donor age to markers of nuclear morphology, nuclear protein expression and markers of heterochromatin, DNA degradation and oxidative stress [MIL 13]. These age-related markers present in aged fibroblasts are lost during reprogramming and do not reappear during differentiation. This indicates that *iPSs do not preserve memories of their age.*

Age-specific markers were induced in fibroblast-derived iPSs and iPSs derived from neurons with Alzheimer's disease, after exposure to progerin. This rapid induction of age-related properties has been used in the modeling of Parkinson's disease. With the induction of iPS aging, disease-specific and age-related phenotypes appeared, such as degeneration of dendrites, neuromelanin expression with age, deregulation of AKT, reduction in the number of tyrosine hydrolase (TH)-positive neurons and mitochondrial ultrastructures of winding and inclusion bodies.

Aging is a new paradigm for iPS cells that can be applied to other cell types to address age-related genetic factors in late-onset diseases.

2.1.3. *A modeling also applicable to epigenetic disorders*

Pathological disorders can result from genetic variants, monogenic or complex. To this must be added the epigenetic disorders directly linked to DNA changes, under the effect of an epigenetic regulation, such as that of methylation. The precise regulation of DNA methylation is little known although it participates in the development of cells and plays an important role in the differentiation of SCs.

Methylation of cytosine in DNA is a major epigenetic factor that regulates several developmental events. Methylation of DNA is carried out by the family of DNA methyltransferases (DNMTs) including DNMT1, DNMT3A and DNMT3B. The mutation of these enzymes is associated with diseases such as cancer, immunodeficiency and neurological diseases. Mutations of *de novo DNA methyltransferase 3B (DNMT3B) cause a genetic disorder called ICF syndrome* type I (immunodeficiency, centromeric region instability and facial anomalies). The deficiency of DNMT3B in patient-derived iPSs induces DNA hypomethylation in the pericentromeric and subtelomeric regions. DNMT3B is very active in human SCs. It is not essential for programming or pluripotency, which makes it possible to study the etiology of ICF type 1 in the context of early development [HUA 14].

Huang's works have shown that DNA methylation plays a role in regulating transcriptional mechanisms and modifying histones to control the expression of cell- or tissue-specific genes [HUA 14]. Upon differentiation, human and murine ESCs undergo a series of *de novo* methylations serving to silence the genes of pluripotency in order to establish differentiated tissue-specific methylations. In the absence of DNMTs, demethylated ESCs undergo apoptosis after the onset of *in vitro* differentiation.

This methylation of DNA also occurs during cell reprogramming. *Epigenic barriers must be overcome to achieve the reprogramming of iPSs* associated with the demethylation of selective pluripotent-gene promoters such as OCT4 and NANOG. Inhibition of DNMTs by 5-aza-cytidine or by inactivation of DNMT1 stimulates partially reprogrammed cells to fully reprogrammed iPSs. In addition, a series of *de novo* methylation occurs during the final stages of reprogramming. These major alterations of the methylome are indispensable during cell reprogramming.

Huang's team generated fibroblast-derived iPSs from patients with ICF1 syndrome, which carries mutations on DNMT3B. The overall profile of methylation by bisulfite genome sequencing has shown that DNMT3B-deficient iPSs show a loss of methylation at the non-CG nucleotides and select the CG hypomethylation of gene promoters. Differentiation of ICF1 iPS reveals hypomethylated CG regions compared to healthy iPSs. These results show the *defects of the methylome and transcriptome in both ICF1-derived iPSs and differentiated somatic cells*, providing an additional argument in favor of the SC systems to study the molecular pathogenesis of the ICF1 syndrome *in vitro*.

One of the aims of the modeling of human diseases is to develop therapies that enable them to be treated and cared for. Animal models are used frequently for research into medicines. For many diseases, however, the animal model does not exist. In addition, these animal models cannot be used for high-throughput screening. This is the reason why pluripotent SC models to be increasingly used in fast and powerful screening techniques.

2.2. A pharmaceutical screening tool

Initially, only laboratories of the pharmaceutical industry could equip themselves with cell culture robots, have a chemical library and work at high throughput. Henceforth, these techniques are also realized in public laboratories. This evolution has enabled a considerable acceleration in fundamental, translational and clinical research, where human SCs offer a wide contribution, thanks to functional genomic screening and pharmacological screening.

Pharmacological screening involves testing the therapeutic potential of a large number of molecules in minimum time, as robots automate procedures and minimize human intervention. The miniaturization of SC cultures enables the high-throughput screening of the pharmaceutical industry; it tests its molecules on microplates comprising 96, 384, or 1,586 wells. Pharmacological screening studies the effect of chemical compounds on the development of normal or pathological cells. This high-throughput screening makes it possible to compare several thousand potentially therapeutic molecules with a relevant biological model of a genetic disease. The discovery of new molecules is thus accelerated. The specific benefit of human ESCs is that they represent a model close to the real physiological framework. We can then have a largely amplifiable cell source, due to the self-renewal ability of ESCs, able to differentiate to a large number of cell types.

2.2.1. *A strategy for approaching high-throughput screening*

High-throughput screening (HTS) can significantly accelerate *the functional exploration of the genome.* It is a platform whose activity consists of modulating the expression of the genes of cells in culture in order to identify the genes involved in the main cell functions and to find potential therapeutic targets. For this, researchers perform transient gene inactivations by tapping into interfering RNA libraries or subjecting the cells to various pharmacological treatments. Then, thanks to the multiparametric microscopy associated with other equipment integrated within an automated chain, they perform high-throughput analyses of any induced cell changes.

High-throughput pharmacological screening (HTS) does not require prior knowledge of the medication, as it can test more than 1 million compounds per study. A limitation exists, however: the phenotype needs to be quantified automatically. This allows the use of cell survival rates or the expression of proteins that can be visualized by fluorescent reporter genes. It is more difficult, however, when the modeled diseases affect a complex phenotype such as electrophysiological defects. In addition, a large number of cells are necessary in order to test a wide range of compounds. This is particularly difficult when the affected cells require a long process of differentiation or when the differentiation percentage is low. The medication candidate approach requires a lower cell count and can be used for complex phenotype alterations.

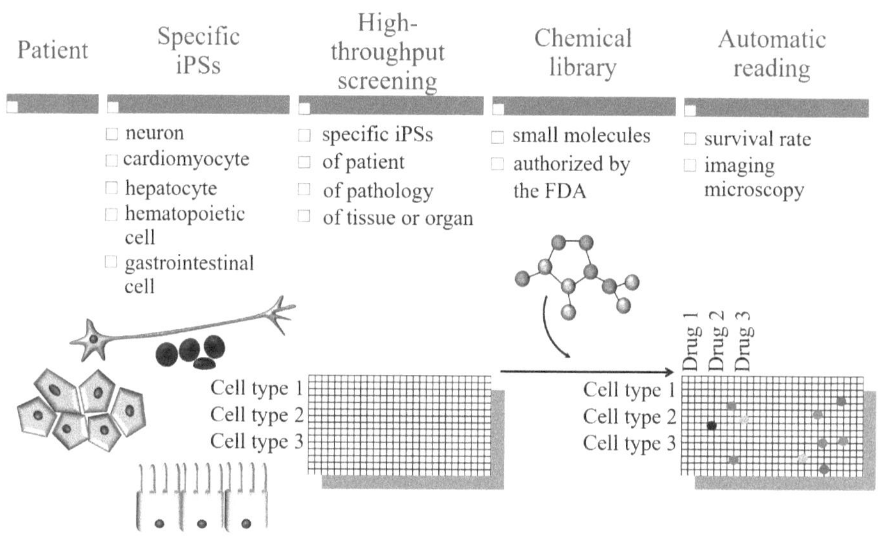

Figure 2.3. *High-throughput pharmaceutical screening. The adult SCs sampled from the patient are reprogrammed into iPSs that can be differentiated into all cell types. Subjected to a bank of chemical molecules, they will be tested on the specialized SCs and enable new drug candidates to be discovered. For a color version of this figure, see www.iste.co.uk/arrighi/stemcells.zip*

After identifying the phenotypes to be treated, the second strategy for identifying the potential medication is *the drug candidate approach*, where a group of well-defined compounds is tested on the affected cells. This methodology is appropriate when an abnormal cell phenotype shows a specific pathway responsible for the phenotype or when similar diseases have already been modeled *in vitro* by these medications. The approach of the medication candidate is the essential part of *personalized medicine*. Thus, as soon as a specific iPS line of a patient is established, the cells can be amplified, differentiated and then used to test a battery of medications with a view of identifying the potential therapy. This approach also concerns the validation of existing medications using new iPS models. The drug candidate approach allows medications authorized by regulatory agencies to be tested immediately on a new disease and thus to quickly bring results before clinical trials on patients.

2.2.2. The search for medications to treat neurological diseases

The drug candidate and high-throughput screening approaches are often combined. For example, Barmada *et al.* first conducted a wide screening of the neurons of diseased mice and then selected a few compounds tested on motor neurons and astrocytes differentiated from iPSs derived from patients with ALS [BAR 14]. It is important to note that these two approaches represent the first step in medication development. Both methods use *in vitro* models and require several phases of rigorous clinical trials prior to therapeutic application.

The cells derived from iPSs have been used to model several types of diseases; it is also the first example of use in the search for medication to treat neurological diseases. This group of diseases includes several types of genetic disorders, early- and late-onset diseases, disorders present in several cell types and disorders affecting both the central nervous system and the peripheral system. In addition, aging (with a doubling of the number of people above 60 years of age in the next 10 years) and the increasing prevalence of neurodegenerative diseases urgently call for new therapies. The evaluation of these degenerative diseases provides a range of uses of iPSs in disease modeling and medication research.

iPS-derived neurons are used as a drug research platform, which began 2 years prior to the discovery of fibroblast reprogramming, in 2009 [LEE 09]. The neural cells derived from iPSs were used for modeling 25 neurological diseases and then for drug screening. Although these diseases are manifested at the end of life, the aberrations of cellular, molecular and electrophysiological phenotypes are detectable in culture. Most studies have been conducted on cells of patients presenting monogenic or chromosomal aberrations, and not on those presenting complex genetic disorders.

Although most cases of Alzheimer's disease and ALS are not monogenic, many models of these diseases rely on fibroblasts from patients carrying mutations on a single gene. In addition, the majority of studies use reprogrammed cells from one or two patients before differentiating them into neurons. This will change when several iPS lineages are available for each disease, in iPS cell biobanks.

Most studies test a restricted number of drug candidates on differentiated cells instead of performing high-throughput screening. Approximately one-third of these studies use iPS-based models to validate existing drugs, rather than researching new molecules. This preference is explained by the rapid transition of *in vitro* models to clinical trials, when these drugs are already authorized by the regulatory agencies. In parallel with potential drugs and high-throughput screening of small molecules, oligonucleotides are also used to target specific sequences. For example, antisense oligonucleotides bind to the mRNA complementary sequences of an aberrant gene and may inactivate it. This methodology is promising because specific gene sequences can be synthesized and *in vivo* pharmacokinetic methods are developed.

2.2.3. *Restoring the expression of the IKBKAP protein in familial dysautonomia*

Several genetic diseases modeled using pluripotent SCs are diagnosed early in childhood. The disease *familial dysautonomia*, which is caused by a mutation in the IKK complex gene (I-κ-B kinase), causes the degeneration of autonomic and sensory neurons, the symptoms of which appear in childhood. Modelings based on iPSs revealed aberrant splicing of the IKBKAP complex [LEE 12]. Drugs correcting this splicing have improved differentiation and neuronal migration.

Familial dysautonomia is a recessive disease caused by a point mutation in the I-κ-B kinase complex-associated protein IKBKAP gene. This mutation leads to *a splicing error that reduces the amount of IKBKAP* or IKAP protein. iPS cell lines were generated from fibroblasts with familial dysautonomia, presenting several disease-specific phenotypes [LEE 12]. Low levels of IKBKAP protein expression were recorded in SCs of the neural crest, which could explain why the disease tends to affect the peripheral system. Familial dysautonomia is a neurodegenerative disease with no effective treatment, hence the search for compounds that would increase the expression of IKBKAP in these affected cells.

Initially, neural cell precursors with familial dysautonomia were isolated by flow cytometry, then amplified and stored by cryopreservation for subsequent screening applications. These precursors presented proliferation, purity and neurogenic differentiation capabilities. The culture conditions in microplates of 384 wells were optimized by a coating method using fibronectin and laminin, allowing good adhesion and improved viability.

The final step of *high-throughput screening* consisted of selecting a disease-specific and sufficiently sensitive criterion, such as an increased expression of IKAP protein in patient cells. A molecular marker kit was developed to quantify the level of transcripts after cell lysis, RNA extraction and finally qRT-PCR, although this technology is not frequently used in high-throughput screening as it directly depends on the amount and quality of the extracted RNA.

Lee *et al.* tested a chemolibrary of 6,912 biologically active compounds of different structures that are authorized by the FDA, that is, a total of 20,736 different treatment conditions. Tested in triplicate versus control wells containing 1% DMSO, the results were normalized with respect to 18S RNA expression. A total of 28 compounds were analyzed as potential candidates to restore the IKAP protein.

After *additional controls of dose-response and cytotoxicity tests* (12 different concentrations), the use of a household gene (GAPDH) and a new validation of the IKBKAP expression in six-well microplates, the following eight candidates were validated: glucosaminic acid (GA); SKF-86466 hydrochloride (SKF); phenindione (PD); dihydrojasmonic acid (DHJ); cyclosporine (CSP); cyproterone (CPT); atropine (AT); 5-Thia-1-azabicyclo (4.2.0) oct-2-ene-2-carboxylic acid, 7-(((2Z)-(2-amino-4-thiazolyl) (methoxyimino) acetyl) amino)-3-(((5-(carboxymethyl)-4-methyl-2-thiazolyl) thio) methyl)-8-oxo-, disodium salt, (6R, 7R) (THIA).

2.2.4. *An example of translational research: ALS*

The modeling of diseases using pluripotent SCs has the primary objective of passing "from the patient's bed to the laboratory". After the sampling of somatic cells from patients, these are reprogrammed as iPSs. Differentiated into numerous cell types, these cells reproduce patient phenotypes *in vitro*. Once the phenotype is characterized, the next step is to look for a drug candidate from a large collection of molecules present in a chemolibrary. Another approach is to test the drug candidates according to the basics that we know of the disease. If the drug is already approved for another disease by the regulatory agencies, clinical trials will be initiated quickly.

Such an approach has recently been used for amyotrophic lateral sclerosis [MCN 15]. ALS is an incurable disease that affects motor neurons. It results from a hyperexcitability at the glutamatergic synapses and the axonal membrane, resulting in a calcium overload, which leads to a degeneration of the motor neurons. ALS is familial in 10% of cases and due to mutations of superoxide dismutase 1 (SOD1) in 20% of cases.

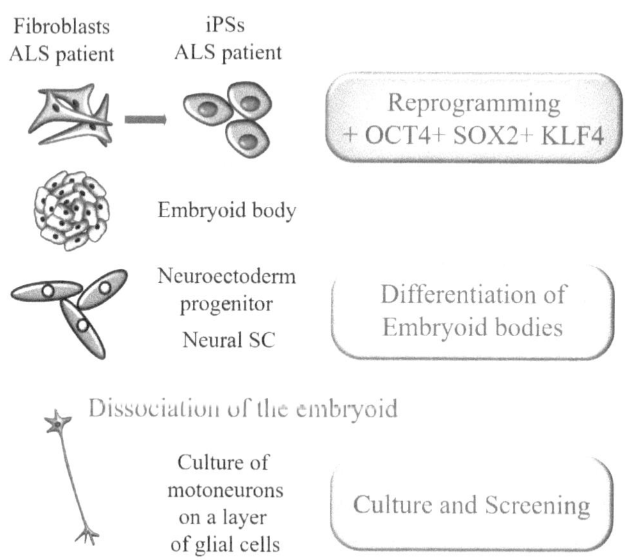

Figure 2.4. *Patient fibroblasts with amyotrophic lateral sclerosis (ALS) are reprogrammed to generate iPSs. iPS-derived motor neurons offer an ideal material for testing treatments for ALS. For a color version of this figure, see www.iste.co.uk/arrighi/stemcells.zip*

The hyperexcitability detected by neurophysiological analyses of ALS patients is found in the motor neurons derived from iPSs of ALS patients presenting a mutation of superoxide dismutase 1 (SOD1). Motor neurons derived from patients' iPS cells also present reduced survival. This phenotype can be corrected by ezogabine, an FDA-approved potassium channel agonist for epileptic patients. The study is then extended to ALS patients carrying mutations on three different genes, namely SOD1, C9orf72 and FUS. In these three cases, the phenotype is restored by ezogabine. As the effects of the drug are demonstrated on several forms of mutations, researchers continue the study *without going through preclinical animal testing* (SOD1 mouse models). The clinical trial should show that ezogabine decreases the excitability of motor neurons in patients by exploring pharmacodynamic markers, two doses compared to a placebo.

As a result, they were able to immediately begin the phase-II clinical trials for this drug aimed at ALS patients. The time interval between the *in vitro* discovery and the approved clinical trial is 2 years, without having to go through animal testing. Treatments for other neurological diseases, including familial dysautomia and spinal muscular atrophy (SMA), are also under clinical development.

To fulfill the promise of iPSs, the transition should now operate in the opposite direction, "from the laboratory to the patient's bed", and should ideally *discover a therapy for patients* who give their somatic cells, which will be used for reprogramming and developing personalized therapies. Several diseases have a wide range of drugs, but not all of them are effective for all patients. Instead of treating a patient with several drugs while waiting to find the right one, all potential medications can be tested simultaneously on the patient's cells, allowing the most effective drug to be prescribed.

2.3. A predictive toxicology tool in the pharmaceutical industry

Toxicity evaluation represents the major obstacle of pharmaceutical research. A large number of drugs (~40%) are abandoned during clinical trials because of the toxic effects that were not detected upstream. It is therefore urgent to improve the effectiveness of predictive toxicology. The regulatory agencies impose on all pharmaceutical research programs that they characterize the *toxicity of the main compounds*. Animal models exist for this, which do not perfectly reproduce the toxicity in humans. These animal experiments are expensive, and the pressure to replace them with alternative approaches is high. *In vitro* studies are increasingly used, especially in standardized preclinical tests based on immortalized cell lines or isolated primary cells. Immortalized cells are genetically modified and may present inappropriate clinical responses. Cells isolated from animal tissues lose their phenotype *in vivo* and present high variability. Differences between species can also influence the results. Most of these disadvantages can be removed by using pluripotent SCs. ESCs and iPSs are of normal karyotype; they can be amplified and are a source of choice for *in vitro* toxicity tests.

The success of reprogramming human somatic cells to iPSs has paved the way for *in vitro predictive models*. iPSs have many properties in common with ESCs, including self-renewal and pluripotency. Access to an *unlimited number of human iPSs* such as cardiomyocytes, hepatocytes and neurons paves the way for genetic and epigenetic experiments that had been impossible to achieve. Extracted from tissues such as skin, blood and other somatic tissues, iPSs can produce different cell lines used during preclinical trials. On the other hand, the possibility of obtaining iPS cells directly from a patient makes it possible to create lines *of iPSs specific to each patient or each disease*. These disease-specific lines are useful for not only toxicity testing but also for understanding the action mechanism of the drug. The potential of patient-specific iPSs would enable the identification of patients who would respond favorably or adversely to the expected treatment; this would be a new approach in personalized medicine.

Predictive toxicology represents a field of application and a strategic benefit for the pharmaceutical industries. Two emerging projects are proof of this. The government and the Association of the British Pharmaceutical Industry have created the Stem Cells for Safer Medicine consortium. Its goal is to develop new *in vitro* toxicity detection tests using pluripotent SCs, tests that can be automated and integrated with the pharmaceutical industry's screening platforms. In the same vein, the aim of the European Stem Cells for Relevant Efficient Extended and Normalized Toxicology project is to provide the proof of concept of pluripotent SC lines in the identification of metabolic pathways.

2.3.1. *Stem cells revealing cardiotoxicity*

Testing the cardiotoxicity of drugs is important as one-third of molecules are withdrawn from the market for this reason. For this purpose, the cardiomyocytes derived from iPSs are exposed to known or unknown drugs for their cardiotoxic potential so as to calculate a predictive index of proarrhythmia corresponding to the possible arrhythmia that this dose of drug could cause.

This cardiotoxicity can be of electrochemical or biochemical origin.

Electrochemical cardiotoxicity occurs when chemical compounds interact with ion channels that generate proarrythmic conditions, which can put the patient's life at risk. This is shown on the electrocardiogram where the QT interval is increased. Pluripotent SCs can differentiate into functional cardiomyocytes, which possess the cardiac electrophysiological properties. Consequently, these SC-derived cardiomyocytes can be used for the screening of compounds on cardiac ion channels.

Biochemical cardiotoxicity is not as well understood as electrophysiological cardiotoxicity. Anthracyclines such as doxorubicin are known cardiotoxic anticancer drugs, the mechanisms of which seem independent of their ability to bind to DNA, probably through the increase of oxygenated free radicals. Kinase inhibitors are also cardiotoxic; they target specific or multiple metabolic pathways present in cancer cells, pathways that are also present in the heart. Cardiomyocytes derived from human iPSs carry out numerous functions of cardiomyocytes *in vivo* and represent an ideal system for evaluating multiple facets of toxicity.

Human cardiomyocytes can serve as a *screening platform*, compared to traditional heterologous expression systems where only one ion channel is overexpressed. The use of human cardiomyocytes avoids trans-species complications. The limitation, however, results from the differentiation protocol, which produces a mixture of several types of cardiomyocytes, mostly cells resembling ventricular cells, as well as atrial and nodal cells.

2.3.1.1. *Evaluation of the cardiotoxicity of medications*

Cardiac toxicity is the leading cause of clinical trials being discontinued. From that point, the medication's development is halted. This is especially true for new molecules that lead to an increase in the QT interval, the time elapsed between depolarization and repolarization of cardiomyocytes. This elongated QT induces ventricular arrhythmias, identified on the electrocardiogram (ECG), in the form of oscillations near the isoelectric line, called "torsades de pointes" (TdP). They can degenerate into ventricular fibrillation leading to cardiocirculatory arrest. The cause of TdPs is the inhibition of hERG (human ether-à-go-go-related gene encoded by KCNH2, potassium voltage-gated channel subfamily H member 2) potassium channel, leading to an increase in the QT interval.

The drug candidates are tested according to the TdP risk that they present. Inhibition of hERG is tested *in vitro* on non-cardiac cell lines overexpressing hERG and recommended by regulatory agencies for all small molecules at the launch of the clinical study. However, not all compounds that inhibit the hERG channel increase QT systematically or induce TdPs. For example, verapamil is an hREG inhibitor used by patients safely, without producing elongated QT or TdPs. Reciprocally, an elongated QT does not necessarily cause TdPs, as proven by ranolazine and alfuzosin, drugs that are considered safe.

Drugs can cause ventricular arrhythmias without inhibiting hERG *in vitro* or prolonging QT in animals. There is therefore a *need to integrate these hERG and QT parameters into a high-throughput screening* that combines the properties of ion channels and cardiac contractions.

The first published high-throughput test using a beating-cardiomyocyte layer from human iPSs was conducted by Guo [GUO 11]. This approach used a 96-well microplate equipped with as many sensors capable of *measuring the real-time impedance* of each cell culture. The impedance between the electrodes reflects the number of cells in each well. By increasing the data acquisition rate, it is possible to detect the contraction of cardiomyocytes in real time and without marking. Human cardiomyocytes derived from iPSs have functional properties of the human heart. After exposure to drugs known for their cardiotoxicity, these cardiomyocytes enabled a ventricular arrhythmia quantification score to be calculated.

Medication screening platforms use automated systems that test thousands of operating protocols on 96- and 384-well microplates. Simultaneously, they follow the culture parameters and the signal detection in fluorescence or luminescence. The system can also be coupled to an electrical field stimulation, thus measuring the oscillations in the intracellular calcium concentration. Automated data acquisition is possible thanks to a high-sensitivity camera and a data-reading frequency of 100 Hz, that is, a measurement interval of 10 ms. In addition, the software analyzes the heartbeat parameters in real time (number of peaks, heartbeat rate and amplitude).

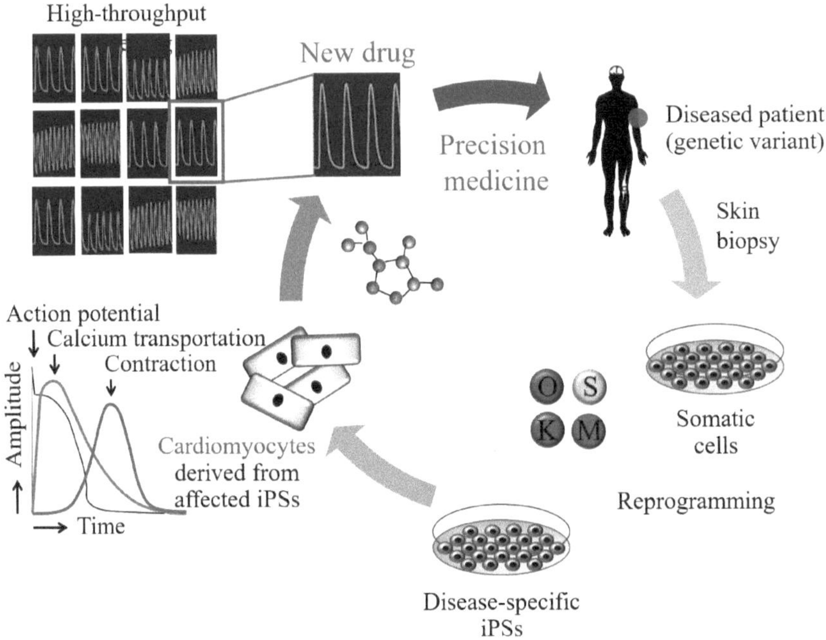

Figure 2.5. *Determination of the cardiotoxicity of drugs using a personalized approach. The patient-specific somatic cells are reprogrammed into iPSs and then differentiated into a cardiac line. The results of* in vitro *cardiotoxicity after drug screening reveal the specific toxicities of the patient and may offer personalized treatment*

2.3.1.2. *The specific case of drugs for cardiac patients*

In order to minimize cardiac risks, the pharmaceutical industries and the U.S. Food and Drug Administration (FDA) require early *in vitro* screening of drug cardiotoxicity. Despite these guidelines, medication-induced TdPs are responsible for many preventable deaths and the withdrawal of pharmacological products from the market.

Cardiac complications are common because preclinical tests have difficulty detecting cardiotoxicity. They are based on the estimation of ion channels expressed by genetically modified animal (Chinese hamster ovary, CHO) or human (human embryonic kidney, HEK) cell lines that do not have the genetic, cellular or biochemical characteristics of human cardiac cells. In addition, channel blocking for a single (individual) ion induces an imperfect QT measurement because the electrophysiology of cardiomyocytes is dependent on often multi-ion (multiple) channels. Consequently, drug screening that blocks an individual channel with CHO or HEK lines can generate false-negative (i.e. alfuzosin) or false-positive (i.e. verapamil) results.

In order to better detect toxicity, preclinical tests should be conducted on human cardiomyocytes, but it is difficult to obtain a patient's cardiac tissue and to amplify his/her cells. With the discovery of embryonic SCs and iPSs, it is now possible to have cardiomyocytes, whose electrical characteristics are the same as primary human cardiomyocytes and which can be amplified in unlimited quantities.

Liang's works have generated a collection of human cardiomyocytes from patients with various heart conditions (LEOPARD syndrome, ventricular polymorphic catecholaminergic tachycardia, hypertrophic familial cardiomyopathy and dilated familial cardiomyopathy), which have been used to model cardiotoxicity [LIA 13]. The duration of action potential and ventricular arrhythmias have been studied at cell level among various cardiomyopathies. These cardiomyocytes have shown sensitivity to cardiotoxic medications, which can predict cardiac complications more accurately than the standard hERG test or screening tests using healthy control lines (ESC or iPS).

Various *cardiomyocyte lines derived from iPSs are marketed* for cardiotoxicity testing. They are part of the cell culture-based platform of tests offering alternatives to animal testing, *called cell-based high-content screening*, HCS, which is necessary for the development of a new generation of safer medications; part of a less expensive strategy.

2.3.2. *Stem cells revealing hepatotoxicity*

SC-derived cardiomyocytes are integrated into drug toxicity tests and are recommended by the regulatory agencies. Other cell types are used in drug safety, such as hepatocytes used in pharmacokinetic testing (ADME tests: absorption, distribution, metabolism and excretion).

One of the many metabolic functions of the liver is the detoxification of endogenous and exogenous substances, including drugs. Thus, several *in vitro* human liver models have been developed from microsomes, cancerous liver cells or *primary hepatocytes*. Primary hepatocyte cultures are considered as the reference for *in vitro* modeling. They can be maintained for several weeks on miniaturized cell culture platforms *in vitro*. However, the scarcity of donors and therefore of healthy liver tissue, inter-donor phenotypic variability and the rapid loss of metabolic function limit the use of primary hepatocytes in high-throughput screening.

With the advancement of SC technologies, different cell types have been produced from iPSs, whose genetic and physiological characteristics are identical to those of the organ and of the pathology from which the SCs originated (affected organ). 2D or 3D culture systems derived from iPSs provide an unlimited source of cells, unlike primary hepatocyte lines. If cardiomyocytes derived from iPSs respond to cardiac function evaluation, then hepatocytes mimic the metabolic capacity of primary hepatocytes. The *cardiotoxicity and hepatotoxicity* depend on various mechanisms, including the formation of free radicals, mitochondrial dysfunction and lipid metabolism disorders. Thus, the simultaneous detection of several phenotypes thanks to the multidimensional combination of high-throughput screening provides *cross-referenced and tissue-specific information*.

Several *in vitro* protocols have developed the differentiation of human hepatocytes derived from iPSs or ESCs. These studies have made it possible to study the metabolism of xenobiotics and to characterize liver detoxification enzymes, such as *cytochromes P450* (or CYPs). The enzymatic profile of hepatocytes derived from iPSs has been explored less because it is more recent.

Recent publications show the predictive sensitivity of hepatocytes derived from iPSs, which distinguish known hepatoxins from compounds whose toxicity is not detected by current methods. iPSs are sensitive to a wide range of chemical compounds, various classes of drugs and toxic mechanisms and generate toxic metabolites in detectable concentrations (aflatoxin B1). The drug detoxification machinery is in place in these hepatocytes derived from iPSs, although in a lower proportion than primary hepatocytes. The major advantage of hepatocytes derived from iPSs, compared to primary hepatocytes, is their *increased viability* and their functionality in culture, enabling exploration of the impact *of exposure to low concentrations for longer durations.* These chronic exposures come close to *in vivo* human exposures to drugs.

Holmgren's works took 14 days following the chronic exposure of hepatocytes derived from iPSs to a panel of drugs (amiodarone, aflatoxin B1 and troglitazone) [HOL 14]. The sensitivity of these cells is proven by EC50 measurements that are of an order of magnitude lower than 48 h of exposure to the HepG2 model cancer lines. A comparison would not have been possible with primary hepatocytes over such a period.

iPS hepatocytes generated from specific patients with idiosyncratic hepatotoxicity will explore the genetic diversity of drug-induced liver damage. The generation of iPS panels reflecting the different cytochrome P450 profiles present in the human population will help advance our understanding of drug metabolism, drug interactions and predictive toxicology models.

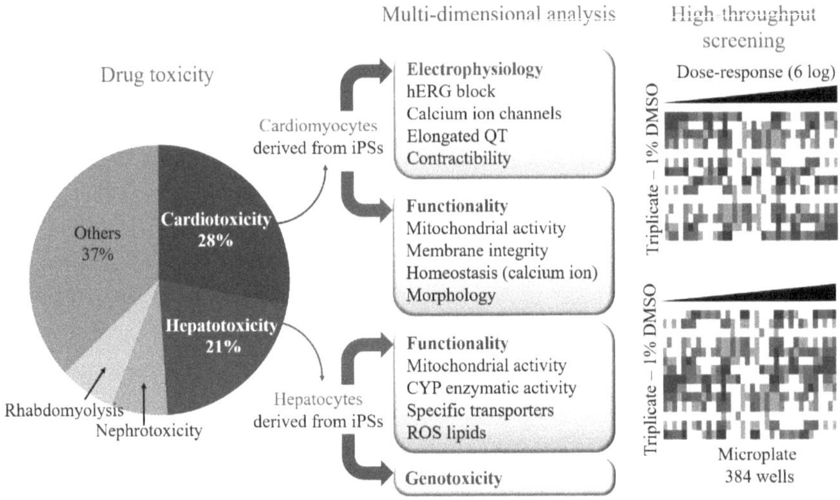

Figure 2.6. *Drug toxicity according to Wilke [WIL 07] and advances in predictive toxicology. For a color version of this figure, see www.iste.co.uk/arrighi/stemcells.zip*

New screening technologies such as *in vitro* tests, *in vivo* models and computer tools continue to be developed to deal with the unwanted events that punctuate the development of drugs. Cardiotoxicity and hepatotoxicity emerge from preclinical trials. The next step is to anticipate this toxicity and detect it early on, as of the initial stages of preclinical development so that these trials can be better transferred to the clinical stages.

3

Stem Cells at the Core of Cell Therapy

"We need to fight to change the world; research forms part of that. Finding new treatments to improve the lives of the patients is one way to do this."

(Marc Peschanski, Director of the Institute for Stem cell Therapy and Exploration of Monogenic diseases)

More than 16,000 scientific publications appeared in 2017 in the field of human cell therapy. Some therapies are validated and commonly used clinically, whereas others are still at the experimental stage. These new treatments paved the way for *cellular drugs*. Cell therapy can be defined as an approach that uses a biological product with therapeutic effect, derived from living cells. In practice, it consists of the injection of human cells with the aim of preventing, treating or mitigating a disease.

Cell therapy is part of the class of *biotherapy*, which aims to produce drugs and therapeutic strategies, based on the living. Their manufacturing and control processes are fully regulated. Their regulation is different from drugs resulting from organic synthesis. Biotherapies include *cell therapies* (stem or differentiated cells), *tissue therapies* (different live tissue transplants), *gene therapies* (gene transfer and intervention on genes) and therapies using drugs that copy natural biomolecules from the human body (growth factors, interleukins and recombinant proteins) and are synthesized by bacteria or cells, such as antibodies or bioactive proteins.

In this chapter, we will follow the progress made in cell therapy: (section 3.1) *Blood SC* grafts, routinely used since the 1970s, enable permanent replacement of a patient's diseased blood system. Thousands of patients benefit from this type of treatment every year. (section 3.2) *Skin grafts*, used routinely since the 1980s, reconstitute layers of new skin in the laboratory and replace that of patients with severe burns. (section 3.3) The new grafts cover several therapeutic areas and extend clinical trials in several countries worldwide.

3.1. Blood stem cells, pioneers of cell therapy

The first bone-marrow transplant was performed by Dr. E. Donnall Thomas in New York in 1956. The patient with leukemia was given a bone-marrow transplant from his identical, healthy twin, hence the concept of *bone-marrow transplant*, because these SCs are found mainly in the pelvis, vertebrae, ribs, sternum and skull. The *blood SCs* are also referred to as *hematopoietic SCs (HSCs)*, hematopoiesis being the process of maturation and differentiation of blood SCs from their bone niche to the release of cells into the bloodstream. HSCs are adult SCs capable of generating circulating blood cells: red blood cells (erythrocytes), white blood cells (leucocytes) and platelets (thrombocytes). Blood cells are vital. They have a short lifespan, ranging from 1 day for some white blood cells to 120 days for red blood cells. Every day, the organism produces hundreds of billions of blood cells in a highly controlled manner, from a small contingent of SCs.

The HSC transplant is not a surgical procedure in the strict sense of the term, but is a complex transfusion, the principle of which is transplantation. It is based on three main actors, namely the source of HSCs (graft), the patient (recipient) and the compatible person (donor). *Compatibility between donor and recipient* is essential. It is defined by the respective histocompatibility antigens. These molecules, present on the surface of the cells of an individual, constitute the major histocompatibility complex (MHC), discovered by Jean Dausset in 1958 [DAU 58]. His works focus on the antibodies of the white blood cells. Unlike the A, B and O blood groups located in the red blood cells, antibodies against white blood cells do not exist in the natural state: they are induced by immunization during a transfusion. Jean Dausset received the Nobel Prize in Medicine in 1980 for his discovery of the *major human histocompatibility complex, or HLA* (human leukocyte antigen).

As soon as a foreign intrusion occurs in the organism, a bacterium, a virus or a parasite, for example, the HLA system detects it and triggers an immune reaction. During an HSC transplant, the recipient's HLA system identifies the donor's HLA system. If they are identical, then there is no immune response and the transplant is accepted. If, on the contrary, a difference is detected, then the recipient's organism develops a defense immune response and the transplant is rejected.

The majority of HSC transplants are carried out from an identical HLA donor within the family: a brother and/or a sister. The probability of compatibility is then 25%. In the absence of a related donor, it is possible to use the SCs of an unrelated compatible donor. There is a significant delay involved in searching for such a donor. To help patients with no compatible donor in their family, *national and international registers of volunteer donors* were established in the 1990s.

– The *related donor* is from the same family as the recipient, with compatible family donors providing the best results. There is donor-recipient compatibility in the tissue group (HLA group).

– The *unrelated donor* is not part of the same family as the recipient. Using the high-resolution HLA typing methods, the results of transplants with unrelated donors approach those made with sibling or other related donors. In general, 50–60% of patients find a compatible donor. Currently, more than 28 million blood and bone-marrow donors and more than 700,000 umbilical cord blood units are listed in international registries. Different sources of HSC are possible, namely bone marrow, peripheral blood and the umbilical cord blood of newborns.

3.1.1. *Bone-marrow blood stem cells*

Bone-marrow removal is performed under general anesthesia by iliac bone punctures. The required amount is 10–20 ml/kg of the weight of the recipient. This type of puncture is replaced by the *removal of SCs from the peripheral blood*. HSCs are harvested after subcutaneous injection of granulocyte colony-stimulating factor (G-CSF). HSCs are mobilized from the bone marrow into the blood and then removed by an extracorporeal circulation (cytapheresis). They are sorted using a cell separator, where red blood cells, white blood cells and platelets are discarded. The rest of the blood comes back through the other arm. The HSC marker is *the CD34 antigen* assayed in blood prior to sampling. The transplant is then frozen in liquid nitrogen. The graft is transplanted to the patient by infusion, after thawing the SCs in a water bath.

Two types of transplants exist, depending on whether the recipient and the donor are two different people (allograft) or the same person (autograft).

3.1.1.1. *Autograft or autologous transplant*

This SC transplantation strategy makes it possible to carry out heavy chemotherapy or chemoradiotherapy, which destroys not only the residual disease but also the bone marrow. The transplant is not a treatment, but a strategy for regenerating destroyed marrow in the case of myeloma, lymphoma and leukemia and sometimes in certain cancerous tumors. In this case, the donor and the receiver are the same person. The autograft process occurs in three stages:

– The *sampling of HSCs* from the patient is carried out at a moment when the tumor volume is as small as possible (complete remission after chemotherapy). The blood graft is frozen or used the same day.

– *Conditioning chemotherapy* is a high-dose chemotherapy that consolidates treatment and has the consequence of destroying the bone marrow. It precedes the injection of the HSCs.

– *Hematopoietic reconstitution or removal of aplasia*: recovery of bone marrow and therefore all blood cells (red blood cells, white blood cells and platelets) after 2–3 weeks. During this period, the patient is particularly vulnerable to infections. Special care is needed, such as regular blood and platelet transfusions.

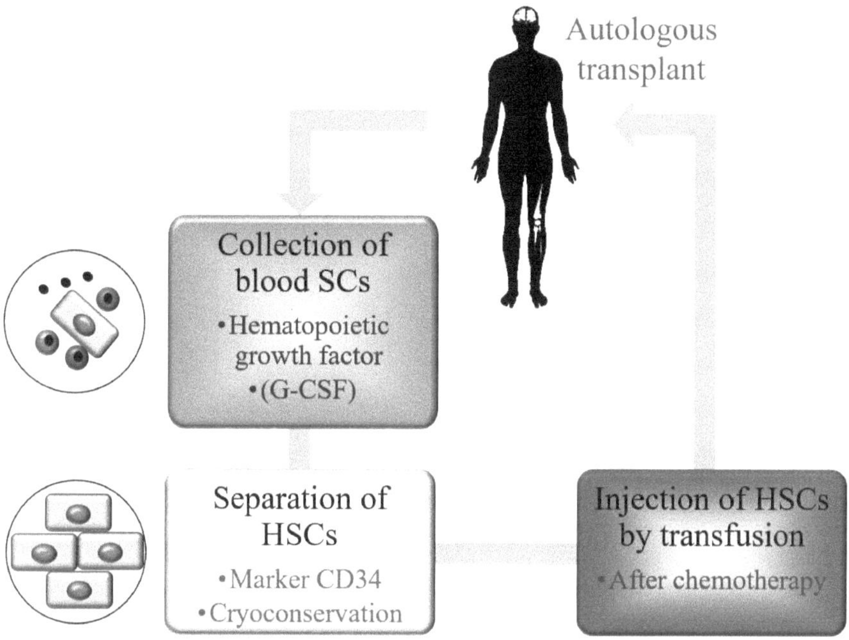

Figure 3.1. *The autologous transplant uses hematopoietic SCs (HSCs) of the patient. They are taken from the bloodstream, then concentrated and frozen. They are reinjected after administration of a high dose of chemotherapy. For a color version of this figure, see www.iste.co.uk/arrighi/stemcells.zip*

The advantage of the autograft is *the absence of rejection* because the patient receives his/her own cells and recovery is faster. Despite the absence of a rejection risk, one of the disadvantages of autografts is that they do not benefit from the grafting effect against leukemia and thus expose the patient to an increased *risk of cancer relapse* compared to allografts. Moreover, the high-intensity treatments administered to the patient in order to eliminate cancer cells resistant to previous therapies result in the destruction of the immune defense system. During this time, the patient is at risk of infectious complications and complications related to the toxicity of chemotherapy and radiotherapy.

3.1.1.2. *Allograft or allogeneic transplant*

If the donor and recipient are different people, the transplant is allogeneic. It is based on the principle of *adoptive immunotherapy*. The donor's immunologically competent cells develop in the recipient's organism. This transplant makes it possible to replace a patient's bone marrow and induce an immune reaction of the transplanted bone marrow against the hemopathy. This immunological effect complements the effects of previous chemotherapies. The allograft is used to treat hematological malignancies such as certain acute leukemias and others that are non-malignant but serious such as immune deficiency, severe thalassemia and aplasia.

After an allograft, the immune system is activated and can trigger two types of reaction:

– *The graft-versus-host disease* (GvHD) response is due to imperfect compatibility between the recipient and donor. This reaction is more common than rejection. It occurs in almost 50% of cases after a familial allograft and in more than 80% of cases after a non-familial transplant. GvHD occurs in acute and chronic forms. (1) In the case of acute GvHD, the organs mainly affected are the skin (rashes), the digestive tract (diarrhea), the liver (jaundice) and the lungs (interstitial lung diseases). The most severe GvHDs that occur in 20–30% of cases carry a risk of death. (2) In the event of chronic GvHDs, any organ may be the target, such as the skin, mucous membranes, liver, eyes, lungs and muscles. Immunosuppressive therapy is routinely prescribed for 6–12 months to prevent GvHD.

– *The graft-versus-leukemia* (*GvL*) effect reduces the risk of relapse. Donor cells (T cells) are able to eliminate residual tumor cells that may be present in the organism. This GvL mechanism is only possible in the case of allografts. Indeed, in the case of an autograft, the patient's immune cells are unable to react against themselves. This effect can be enhanced further by donor T-cell transfusion, an approach that in some cases may lead to a complete new remission if the disease has relapsed despite transplantation. In this case, however, it is necessary to take into account that the risk of acute GvHD is not negligible.

A new HSC transplant technique, the mini-graft, reduces the side effects of conditioning and the duration of the hospital stay. It associates the GvL effect of the lymphocytes contained in the donor graft with respect to the cancerous cells of the recipient. HSCs are transplanted after chemotherapy and/or reduced radiotherapy, supplemented by immunosuppressive therapy. These mini-grafts are better tolerated, allowing older patients and/or those who are less physically fit to benefit from an allograft.

To summarize in a few figures, the medical and scientific report on collection and transplantation in France, an annual report presented by the Agency of Biomedicine, lists *1,964 HSC allografts* (bone marrow, peripheral blood or cord blood) carried out *in 2015* on 1,936 patients [AGE 15]. In 2015, *3,123 autografts* were performed on 3,007 patients. In 99.6% of cases, the graft consists of HSCs from peripheral blood. The distribution according to the origins of the SCs shows that a majority of SCs come from peripheral blood (67%), followed by SCs from bone marrow (27%) and finally those from cord blood (6%).

In Europe, HSC transplantation is a frequently used treatment (in 2008: 27,000 transplant patients, including *11,000 by allogeneic transplant and 16,000 by autologous transplant*) for benign and malignant hemopathies and pediatric as well as metabolic and immune disorders. The discovery of SCs in fetal-placental blood made a third source of HSCs available.

3.1.2. *Blood stem cells from fetoplacental blood*

HSCs of fetoplacental blood or *SCs of cord blood* are present in the umbilical cord and in the placenta during delivery. The collection of cord blood is safe for both the mother and the newborn. These SCs do not correspond to fetal totipotent SCs; they are *multipotent* and capable of reconstituting, during a transplant, the entire hematopoiesis in the recipient, that is to say, the circulating blood cells. These HSCs are more immature than those found in adult bone marrow. However, this immaturity does not make it possible to overcome the HLA barrier and there must be compatibility between the recipient and the cord cells.

The *first HSC transplant from cord blood* was performed in 1988 in the Hematology-Marrow Transplantation department of Professor Eliane Gluckman at the Saint-Louis Hospital in Paris [GLU 89]. This transplant was performed on a child who had aplastic anemia due to Fanconi disease. The cord blood cells that were used for transplantation had been removed and frozen at the time of the birth of his younger brother. Since then, more than 10,000 patients worldwide have received cord blood transplant.

3.1.2.1. *Collecting blood stem cells from fetoplacental blood*

Before the birth of the newborn, the mother completes a medical fitness questionnaire used in all blood transfusion centers. It comprises several questions regarding her health, lifestyle and travels in order to assess whether she is fit to donate. This information is confidential and subject to medical secrecy.

At birth, the fetoplacental blood is collected after the umbilical cord is cut. The absence of infectious risk is controlled. The SCs are separated by the specific surface marker of type CD34 and HLA. They are cryopreserved in liquid nitrogen in a barcode-anonymized pocket stored in a biobank, which gathers the *sample data (HSC) and clinical data from the mother (donor)*.

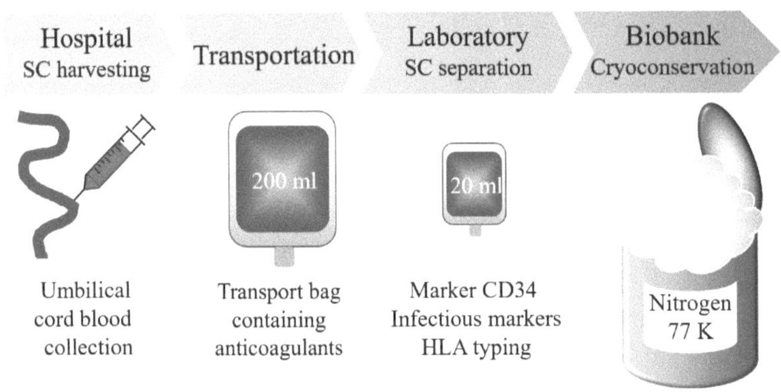

Figure 3.2. *From the hospital to the biobank, the placental blood SCs are collected, sorted and then stored in order to ensure the traceability and availability of the SCs, managed by national and international registries*

3.1.2.2. *The umbilical cord, a precious source of blood stem cells*

When frozen at the birth of the newborn, fetoplacental HSCs present advantages compared to the HSCs of an adult donor:

– they are *available more quickly* as they are already stored in biobanks;

– their capacity for *proliferation is increased*;

– their *immunological immaturity* induces less rejection and hence HLA incompatibilities are more easily tolerated;

– the *compatibility* of fetal and placental HSCs can offer a therapeutic solution to patients who do not have a compatible adult donor by high-resolution HLA typing.

However, it should be mentioned that the major limitation of this fetoplacental HSC transplant is the *reduced amount of HSCs in the umbilical cord*, because the blood volume of the order of 200 ml remains low. This explains why these transplants are reserved for *pediatric oncology*, where the number of HSCs required is less than that of a transplant in adults. In pediatrics, the mother of a child with a hematological malignancy who is expecting a second baby may decide to have HSCs taken from the newborn. They will be banked and dedicated to her affected child. By way of derogation, there may be a retention for intra-family use if a brother or sister has hematological disease requiring an HSC transplant and if a compatibility exists between the placental blood cells and this brother or sister.

Regarding the autologous transplant, there is currently no evidence as to its efficacy and safety in the treatment of malignant diseases. Autologous transplantation does not provide the immune cells capable of eliminating leukemic cells like an allogeneic transplant does. Concerning the treatment of cancers, the HSCs of the patient's peripheral blood or bone marrow, collected in greater quantity, are favored over cord blood.

Throughout the world, more than 600,000 units of placental blood are referenced in more than 100 banks available to transplant physicians (bone marrow donors worldwide). The goal of biobanks is to guarantee quality units, for which *the analysis, qualification and conservation of the graft are validated*. The priority is to favor units extremely rich in SCs. These banks work in a network and supply transplant centers worldwide.

3.2. Skin stem cells

Skin reconstruction largely remains at the research and development stage. The idea would be to be able to grow on demand, in the laboratory, autologous skin, composed of both a dermis, for its mechanical properties, and an epidermis, for its barrier function. The first works date from the late 1970s; they paved the way for the creation of artificial skin by the development of a dermal equivalent composed of fibroblasts and collagen, with this dermal equivalent then covered by keratinocytes; the essential cellular actors of the skin. On this complex architecture, clinical trials now invest the functional properties, which are based on the structured tissues of the skin.

3.2.1. *Structure of the skin*

The skin is a complex organ providing protection, sensitivity and thermoregulation functions. It covers approximately 2 m^2 of the body surface and contributes to 15% of the weight of an individual. It consists of *three superimposed tissues*: the epidermis, the dermis and the hypodermis. The first two correspond to the cutaneous tissue, whereas the last one represents the subcutaneous tissue.

– The *dermis*, of mesenchymal origin, contains *fibroblasts* responsible for the synthesis and remodeling of the extracellular matrix of the skin, composed of collagen, elastin, proteoglycans and hyaluronic acid. The dermis is crossed by capillaries and nerve endings that ensure its nutrition and innervation as well as those of the underlying epidermis.

– *The epidermis*, of ectodermal origin, is composed of *keratinocytes* that are constantly renewed from the dermis. In their maturation process, keratinocytes migrate from the dermis to the skin surface and transform into corneocytes; anucleate flat cells that form an impermeable physical barrier, the stratum corneum. The epidermis also includes cells from the neural crest, the *melanocytes*, responsible for *photoprotection*. It also includes *macrophages* from bone marrow and *Langerhans cells*, which are dendritic cells presenting antigens of the organism, and therefore of *immunoprotection*. The epidermis is also at the origin of adnexal formations, veritable epidermal buds penetrating into the dermis, which are the sweat glands, sebaceous glands and hairs.

– *The hypodermis* is a connective tissue containing *adipocytes*, a place for the storage of lipids which have the dual role of thermal insulator and energy reserve. It is very richly *vascularized and innervated.* The hypoderm allows for the regulation of cellular metabolism. It gives the skin mechanical protection and elasticity.

The dermoepidermal junction is a docking area between the dermis and the epidermis. This interface is a metabolic exchange area allowing selective filtration between the dermis and the epidermis. Its wave-shaped structure is characteristic. It plays an important role in reepidermization during healing because, through the glycoproteins that constitute it, it serves as a support for the adhesion and migration of keratinocytes. It is noted that this junction is composed of the plasma membrane of the basal keratinocytes as well as its attachment structures or hemidesmosomes, the *lamina lucida* traversed by the anchoring filaments, the *lamina densa* composed of collagen, and the fibrillar area composed of anchoring fibrils and dermal microfibrils.

3.2.2. *Reconstruction of a simplified human skin*

It all started in 1978 in Boston, when E. Bell's team at MIT cultured fibroblasts in the presence of collagen and observed that the collagen matrix was contracted by fibroblasts. This discovery is at the origin of a series of works intended to establish whether the fibroblasts cultivated *in vitro* come close to those of human skin. In 1979, Bell *et al.* reconstituted the first *living dermal equivalent* by seeding a mixture of fibroblasts and type I collagen [BEL 79]. By developing long cellular extensions, fibroblasts induce the organization of the collagen matrix into a resistant disk. Subsequent studies have shown that when fibroblasts are cultured in the dermal equivalent, they regain physiological behavior. Under these conditions, the fibroblasts synthesize a mature collagen possessing the telopeptides necessary for the polymerization and cross-linking of collagen fibers [NUS 84], which fibroblasts cultured in monolayer are incapable of producing.

In order to cover the dermal equivalent with an outer layer or epidermis, hence the term *epidermization of the dermal equivalent*, keratinocytes, previously isolated from the tissue by enzymatic extraction, are seeded directly on the dermal equivalent and cultured at the medium–air interface.

An *equivalent skin* presenting a well-differentiated and organized epidermis develops. Another method consists of depositing a small epidermal biopsy in the center of a dermal equivalent; epidermization then occurs step by step from the explant [COU 86].

The principle of reconstructing the human skin is to put the cells back into a physiological environment. There are several procedures for reconstructing an entire skin, the aim of which is to provide a study model of the biology of the skin and especially to assess the possibilities of transplants in humans.

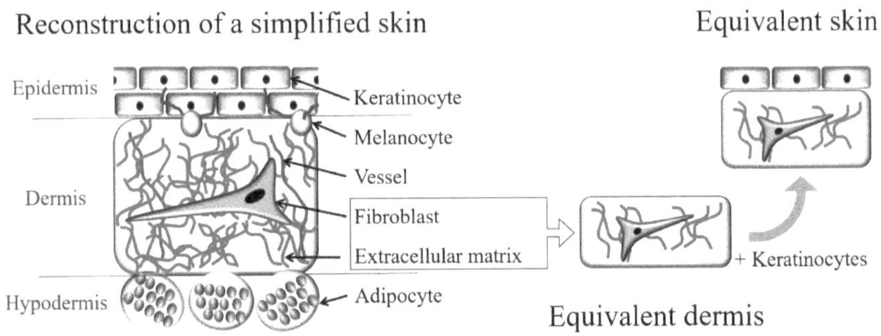

Figure 3.3. In vitro *reconstruction of a simplified skin. The dermal equivalent allows cell–matrix interactions. The equivalent skin restores tissue interactions. For a color version of this figure, see www.iste.co.uk/arrighi/stemcells.zip*

Louis Dubertret's works showed that cell differentiation, in these *in vitro* organogenesis models, is similar to that which takes place *in vivo*. They have enabled the action of different drugs to be studied, *ex vivo* pigmentation techniques and *in vivo* healing models to be developed and some of the anomalies involved in the pathophysiology of psoriasis to be rebuilt in three-dimensional culture. This technique enabled, in 1983, the *world's first grafts of completely reconstituted human skin* – dermis and epidermis – in approximately 10 children suffering from giant cutaneous malformations.

3.2.3. *Constantly evolving skin reconstruction*

Healing processes are characterized by the formation of granulation tissues accompanying the contraction of these tissues. These processes involve specific mesenchymal cells of the scar tissues, the *myofibroblasts*. They possess morphological and metabolic characteristics that are intermediate between those of fibroblasts and those of smooth muscle cells. They express the alpha-smooth muscle actin (α-SMA) specific to smooth muscle cells. This actin isoform suggests that these cells have acquired contractile properties responsible for the phenomenon of cicatricial contraction. Two concomitant principles explain the contraction processes:

– the *migration of the fibroblasts* from the margins to the center of the wound provokes the reorganization of the extracellular matrix and the contraction of the tissue;

– *myofibroblastic differentiation* could induce tissue contraction. Indeed, myofibroblasts exert mechanical tensile forces greater than those exerted by the fibroblasts.

Different *in vitro* models have been used to decipher the contraction mechanisms. Monolayer culture enabled partial study of the phenotypes and the proliferative potential of the cells, without having access to the functional properties. By contrast, the three-dimensional fibroblast culture models in a collagen or fibrin matrix have been able to serve as study models for wound healing [GRI 94]. Indeed, studies carried out by cinemicrography have shown the establishment of dynamic interactions between the fibroblasts and the collagen matrix [BEL 79]. The composition of the matrix and its concentration play an essential role in the contraction processes *in vitro* [GIL 86].

These studies have highlighted the prominent role of myofibroblasts. They show an increased ability to induce *collagen matrix contraction*. Finally, the presence of fibroblasts in the collagen matrix and the phenomenon of matrix reorganization enable *better keratinocyte growth*. Fibroblasts secrete growth factors that are important for epidermal development. These growth factors are diffusible products that can act remotely, without direct contact between the cells [COU 86]. The *keratinocyte growth factor* (KGF) is synthesized by the fibroblasts, while the membrane receptor needed for KGF signal transduction is specifically expressed by the keratinocytes of the basal lamina [RUB 95]. When fibroblasts and keratinocytes are mixed, the rate of contraction is amplified, which shows a phenomenon of cellular cooperation: the keratinocytes and fibroblasts act in synergy.

3.2.3.1. *The model of skin reconstructed* in vitro

The model of reconstructed epidermis marketed by SkinEthic (Episkin group of L'Oréal) is obtained by culturing keratinocytes on a synthetic support of the polycarbonate membrane type [ROS 90]. Other reconstructed epidermis using this principle have also been developed: *EpiDerm* from MatTek and *Mimetop* from BASF Beauty Care. Finally, other models consist of covering an inert synthetic support with a biological support (collagen gel) so as to promote adhesion, proliferation and differentiation of the keratinocytes. This is the case for L'Oréal's *Episkin* model [TIN 91]. These models can be supplemented by integrating other cell types such as melanocytes [TOD 93] or Langerhans cells [REG 97].

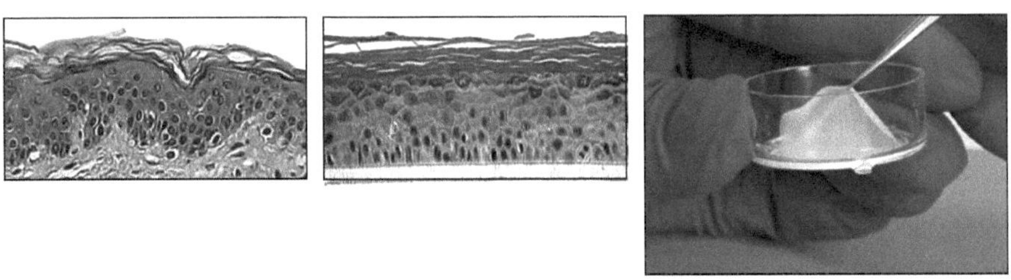

Figure 3.4. *Histological section of epidermis* in vivo *and* in vitro. *Photo credit: Philippe Gotteland, http://www.episkin.com. For a color version of this figure, see www.iste.co.uk/arrighi/stemcells.zip*

The first *reconstructed dermis* was the lattice of collagen [BEL 79], where the fibroblasts reshape the collagen matrix by contracting it. The lattice retracts and thus forms a dermal equivalent. With *collagen sponges*, collagen-based biopolymers, retraction phenomena are not observed. The Tissue and Cell Bank model by Dr. Damour is composed of type I and III collagen, chondroitin-4-sulfate and chondroitin-6-sulfate as well as chitosan [SHA 90]. This support is marketed by BASF Beauty Care under the name

Mimedisk. After seeding, the fibroblasts will reshape the sponge and produce extracellular matrix (ECM) to give a dermal equivalent, *Mimederm* [BER 97].

Finally, only a few models consist of letting the fibroblasts neosynthesize the ECM without preexisting matrix support. These models are based on the use of ascorbic acid to promote fibroblast proliferation and the production of ECM [PHI 94]. These equivalent dermas can be endothelialized by the addition of umbilical endothelial cells in lattices [HUD 03] or collagen sponges [TRE 05].

3.2.3.2. *Techniques used in the human clinic*

From skin biopsies performed on the patient, if possible in a hair area in order to have SCs, it is possible in 3 weeks to obtain 1 m^2 of epidermis from a biopsy of 2 cm^2. This epidermis is formed of keratinocytes in culture on a layer of fibroblasts, according to the technique of Rheinwald and Green. Fibroblasts are cultured until they form a confluent monolayer [GRE 77]. They are then irradiated in order to stop proliferation while maintaining cell survival. This feeder layer is then seeded with epidermal cells. The very fragile epidermal layers obtained are stapled onto gauze compresses that enable them to be handled. A fibrin support is under study. Protected by patents and commercialized on an industrial scale by the American firm *Genzyme Tissue Repair* of Boston (formerly BioSurface Technologies), this technique makes it possible to reconstitute the protective barrier of an autologous epidermis in burn victims. However, its poor trophic and esthetic qualities and its prohibitive cost reserve it for the treatment of severe burn patients (60% of the body surface covered by third-degree burns). Although it is possible to obtain cultured autologous epidermis quick enough for clinical use, the ideal support of dermal equivalent for grafting still remains to be found.

Current possibilities include:

– *The cryopreserved allograft of cadaver skin*, which allows the burned patient to be covered first, with the epidermis being amplified in the laboratory. Once removed by abrading the epidermal antigen of the corpse, this underlayer will be used for the transplantation of the culture epidermis.

– *Integra* (Integra Life Sciences) is an *acellular collagen matrix* derived from bovine tendons and shark cartilage proteoglycans, coated with a silicone sheet. This matrix is progressively vascularized and colonized by the patient's fibroblasts, which replace the matrix collagen with autologous collagen. After 3 weeks, when adequate vascularization of the dermal matrix has been obtained, the silicone sheet is removed and replaced with an autologous thin skin graft. As regards autologous epidermis grafting on this type of product, it still poses some unresolved problems.

– *The Appligraft* by Novartis is a culture of *allogeneic fibroblasts in a collagen network*, resulting in the formation of a dermal equivalent onto which an autologous keratinocyte culture can be grafted. Current research favors allogeneic transplants (keratinocytes and fibroblasts) by combining the principles of cell therapy and cell culture and by going far beyond the rare indications of large burned surfaces to focus on small surfaces of chronic wounds, which are much more frequent.

3.2.3.3. *Toward a reconstructed total skin*

For severe burn patients, it is possible to *generate a functional epidermis in vitro* that, grafted to the patient, enables restoration of an epithelial barrier essential to the patient's survival. A small fragment of healthy skin measuring a few square centimeters is cultivated for several weeks on a large scale in the presence of culture media and growth factors to generate several square meters of skin. In this case, only the epidermis is restored, which gives the skin high fragility. These cellular biotherapies of epidermis generation are developed by certain private companies and are expensive. One solution for the future is to generate these skin cells from pluripotent, embryonic or induced SCs and to reconstitute both the epidermis and the dermis.

Let us note that despite attempts made, keratinocyte cultures do not "take" on the dermal equivalent because no dermoepidermal junction reconstitutes in the case of cultured keratinocytes placed on a revascularized dermal equivalent. Indeed, in the case of the dermal equivalent, it is an autograft (and not an epidermal cell culture) that is performed on the revascularized dermal equivalent. As a result, research is moving toward a *reconstructed total skin model*, seeded with SCs that, by differentiating, will be able to recreate different cell layers of the skin, and especially the basement membrane. Another avenue of biotechnology research also focuses on *growth factors*, produced by mobile cells (fibroblasts, platelets and histiocytes). However, to be effective, these growth factors must be not only used in enormous concentrations (relative to their natural state), but also associated with each other, hence the alternative *of bringing the cells that secrete all these factors*. Besides the exposed techniques of autografts, homografts, cell cultures (alone or associated with cadaver skin dermis) and dermal equivalent, a "devitalized" human dermis exists, without cells, which is then seeded with fibroblasts. In France, also under study (in phase II) is a *reconstructed total skin* that may be available within 5 years. It is an extracellular matrix in which fibroblasts, SCs or even fat cells from the donor are placed and cultured.

3.3. Stem cells at the core of clinical research

Blood transplants and skin grafts have paved the way for SC transplants. They have been tried and tested, and certain cell therapy drugs have been granted marketing authorization, such as Holoclar, a treatment involved in regenerative medicine of the eye. Moreover, it is in the indication of the treatment of pathologies such as age-related macular degeneration (AMD) that the cell therapy projects are the most advanced. Drugs derived from cell therapy form part of advanced therapy medicinal products, alongside gene therapy medicinal products, those derived from cell or tissue engineering and those resulting from a combination of the three types mentioned. However, most are still in the clinical trial stage, using embryonic SCs and the induced pluripotent SCs. They are investing in new therapeutic areas and recruiting clinical investigation centers worldwide.

3.3.1. *State of progress of global-scale clinical trials (January 2017)*

Access to clinical studies has been facilitated by the Internet. Veritable platforms communicate on ongoing studies, future protocols and the results obtained. These information sites are aimed at patients, healthcare staff and volunteers wishing to participate in clinical trials alike. The Clinical Trial Registry of the US National Institutes of Health (NIH) lists all studies by clinical trial goal, patient profile and study center (which exist in more than 200 countries (https://www.clinicaltrials.gov/)).

As of early January 2017, the clinical trials included 960 ongoing SC transplant studies. Virtually all of these trials use *mesenchymal SCs*. These tests mainly explore diseases of the *immune system* (441 trials) and *hematological diseases* (335 trials). Three trials concern *ESCs*, with two clinical trials on macular degeneration and one on cardiac ischemia. None of the trials concern iPSs.

The majority of the tests are at the very beginning of the process, *in phase I* (361 trials) *or in phase II* (369 trials). There are only 75 trials in phase III and 28 trials in phase IV. Of these 28 trials in phase IV, the two major families of diseases studied are blood diseases (10 studies) and those of the immune system (10 studies). Approximately a quarter of these trials are subsidized by the industry (220 studies) and 137 studies by the US National Institutes of Health (*NIH)* and US government institutions. Concerning SC-based clinical trials, the majority of trials take place in the United States and in Europe, mainly in France and Germany.

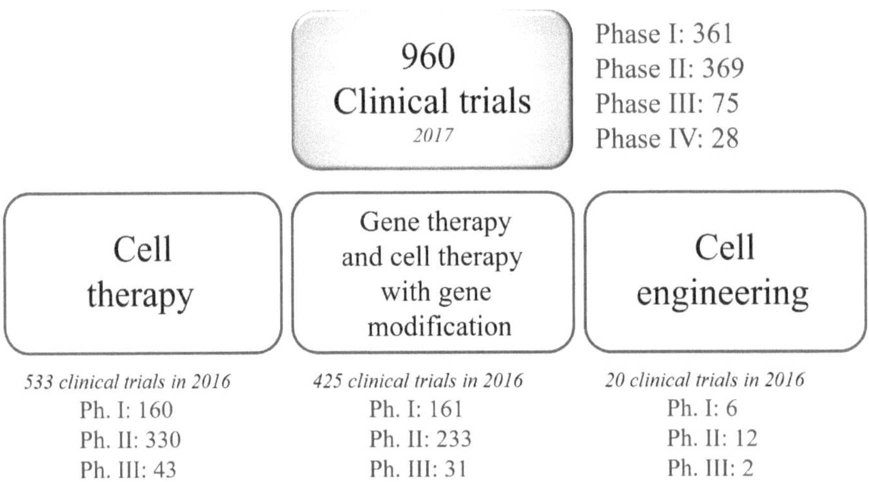

Figure 3.5. *Number of ongoing clinical trials on SCs in phases I, II, III and IV according to the type of disease as of early January 2017 – clinicaltrials.gov*

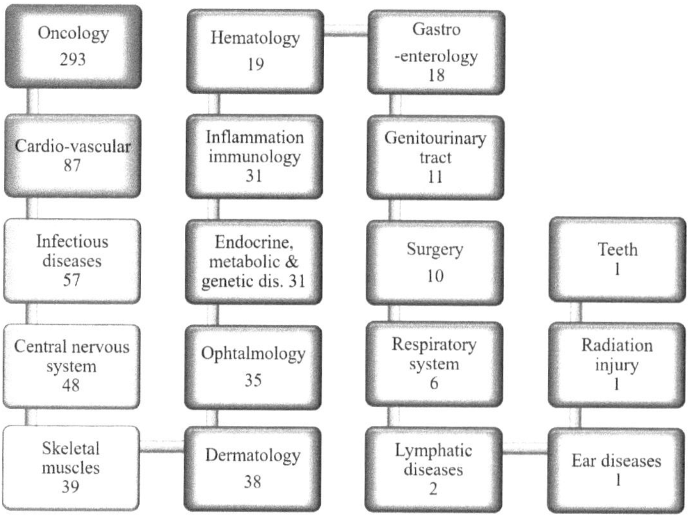

Figure 3.6. *Number of clinical trials on SCs according to pathologies in early January 2016 according to the clinicaltrials.gov website. For a color version of this figure, see www.iste.co.uk/arrighi/stemcells.zip*

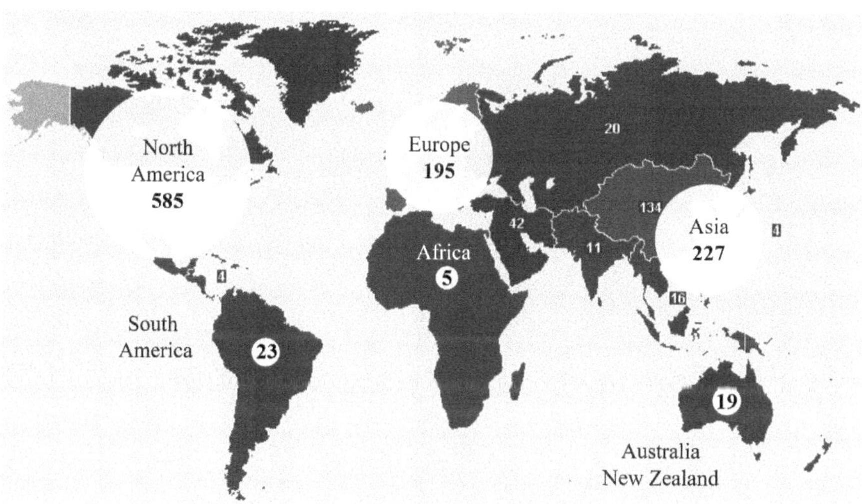

Figure 3.7. *Number of clinical trials on SCs worldwide according to the clinicaltrials.gov website in February 2017*

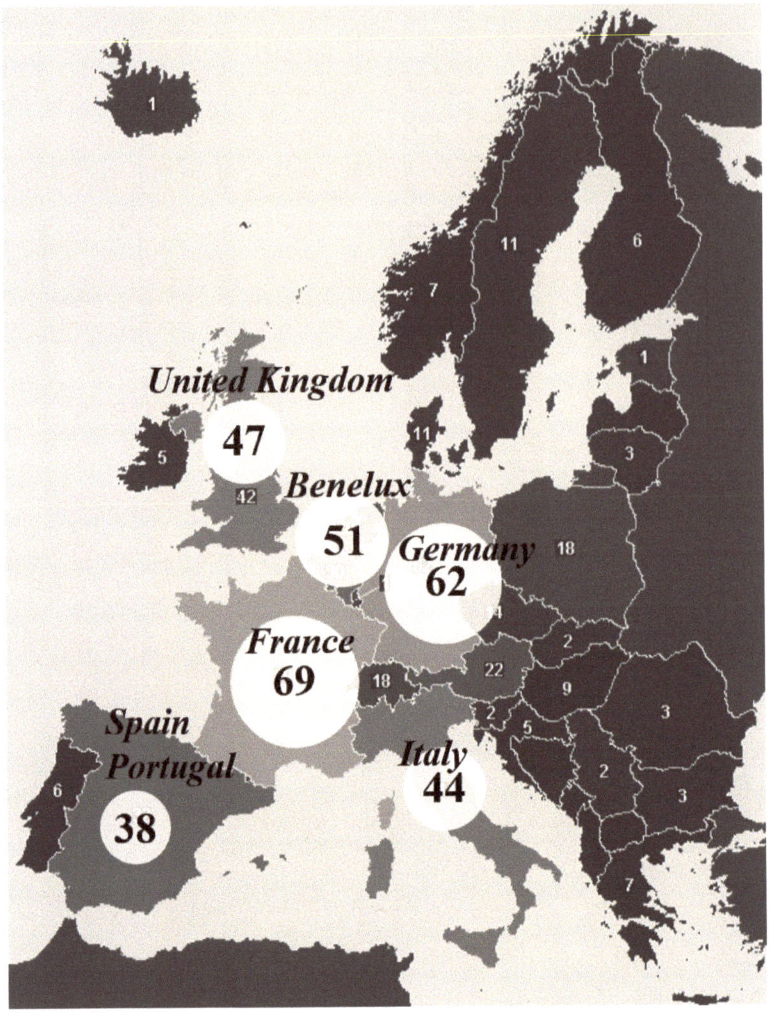

Figure 3.8. *Number of clinical trials on SCs in Europe according to the clinicaltrials.gov website in February 2017*

3.3.2. *The first cell therapy medicinal product to be licensed in Europe, Holoclar*

In 2014, the Committee for Medicinal Products for Human Use (*CHMP*) of the European Medicines Agency (EMA) authorized the sale of Holoclar, developed by the Italian company Holostem [EMA 14]. This is a treatment linked to *regenerative medicine of the eye*. This is the first innovative therapy containing SCs, indicated for the treatment of moderate to severe keratitis due to limbic cell deficiency. This deficit affects approximately 3 in 100,000 individuals in the European Union; it is a rare disease. Used in the treatment of a rare disease, Holoclar has been designated an orphan drug.

Holoclar is an SC-based treatment used to replace damaged cells on the surface of the cornea, the transparent layer located at the front of the eye and covering the iris, which is the colored part. It is used in adult patients suffering from a limbic SC deficiency caused by chemical eye burns. Patients suffering from this disorder do not have enough limbic SCs, the role of which is to renew the outer cells of the cornea. Holoclar consists of removing the patient's limb cells from the edge of the cornea, amplifying them and then transplanting them to the damaged corneal surface.

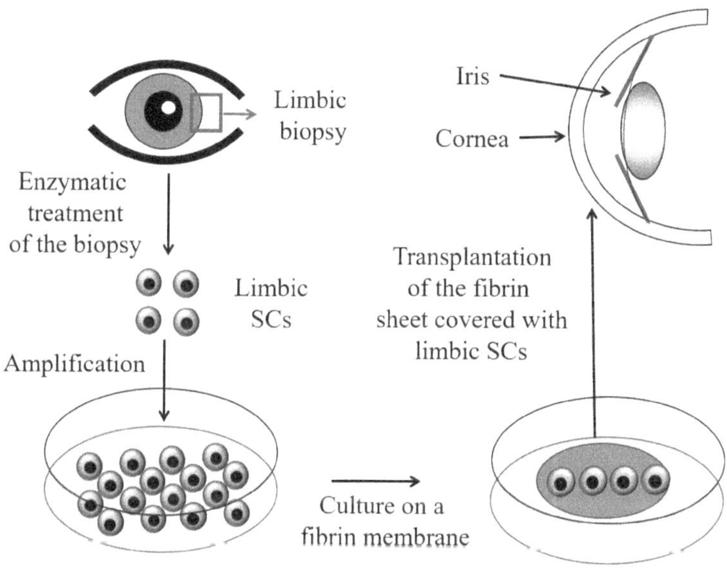

Figure 3.9. *Holoclar is an SC-based treatment used to replace damaged cells on the surface of the cornea. For a color version of this figure, see www.iste.co.uk/arrighi/stemcells.zip*

Holoclar consists of a transparent circular sheet of 300,000–1,200,000 viable autologous human corneal epithelial cells. This medicinal product is the result of an industrial process consisting of amplifying the SCs *ex vivo* from a biopsy sample of 1–2 mm^2 of undamaged limb made in the patient who will receive the cellular sheet. The therapeutic goal is the replacement of the corneal epithelium and limbic SCs. During the corneal repair process, the implanted SCs are intended on the one hand to multiply, differentiate and migrate in order to regenerate the corneal epithelium and, on the other hand, to maintain a reservoir of SCs able to continuously regenerate the epithelium of the cornea. Each preparation contains an individual dose of treatment with a sufficient number of cells to cover the entire surface of the cornea. The recommended dose is 80,000–300,000 cells/cm^2, corresponding to 1 cm^2 of product/cm^2 of lesion. Each preparation is intended for a single treatment.

Biopsic limb samples are taken under local anesthesia. After washing of the ocular surface with a sterile balanced salt solution, the conjunctiva is detached from the limbus to expose the corneal site of the sampling. A 2 × 2 mm incision is made for the biopsy, which is received by the manufacturer within 24 h of collection. A topical antibiotic prophylaxis is administered after collection.

Holoclar is exclusively intended for the regeneration of autologous limbic SCs, in compliance with the approved therapeutic indication. It needs to be implanted under aseptic conditions together with a limb peritomy, a conjunctival incision and an excision of the corneal fibrovascular tissue from the defective area. Next, the insert is put in place under the incised conjunctiva. The excess insert is cut away, and the edges are covered by the conjunctiva and sewn with three stitches in order to form a physical seal of the lesion and secure the implant. A topical, systemic anti-inflammatory and antibiotic prophylactic treatment is administered after implantation.

3.3.3. *Ongoing clinical trials using human embryonic stem cells*

Adult SCs have enabled validation of the regeneration of the limbic cells of the eye. Other therapeutic strategies have focused on ESCs with, firstly, Geron, which plans to treat spinal cord injuries by oligodendrocyte injection. Despite having entered phase I, the clinical trial will be discontinued. Next follows Advanced Cell Technology, whose clinical trial is conclusive.

3.3.3.1. *The world's first spinal cord injury study, launched by Geron*

In January 2009, American biotech company Geron (Menlo Park, California) got approval from the US Food and Drug Administration (FDA) to begin the first clinical trial based on human ESCs (hESCs). The product *GRNOPC1 is the first therapeutic candidate* containing *oligodendrocyte progenitor cells* (OPCs) derived from hESCs. These progenitors stimulate the growth of nerve tissue and its myelination. The FDA's authorization allows Geron to launch the clinical study on 10 patients suffering from *acute lesions of the spinal cord*, aiming to restore them through oligodendrocyte progenitor cell injections [KEI 05]. The treatment administered 15 days after the lesion was to enable regeneration of the myelin sheath.

The oligodendrocyte progenitor cells of the GRNOPC1 line are capable of differentiating into oligodendrocytes, belonging to the glial cell family. They play a major role in the transmission of nerve impulses, by emitting membrane extensions wrapped around the axons and covering them with myelin. Without this insulating sheath, the neurons can no longer transmit the electrical impulse. Moreover, oligodendrocytes secrete nerve growth factors (NGF), whose stopping leads to neuronal degeneration. The idea is to inject, at the lesion area, cells able to differentiate into new oligodendrocytes allowing neurons to regenerate.

Preclinical studies have shown that the GRNOPC1 injection restores locomotion in rats with a spinal cord injury [KEI 05]. The artificial lesions created in rats reproduce the same clinical signs as in humans, that is to say, a loss of muscle function in the trunk and the hind legs as well as a loss of bladder control. The animals received injections of GRNOPC1 at the spinal lesion 7 days after the surgical procedure that reproduces the trauma. In comparison with the control group, the animals injected with SCs presented a *significant recovery in the locomotor activity* of their legs and an increase in stride length. Histological analysis of the spinal cord at the injections showed an *increase in the remyelination* and the number of axons near to the injection areas. The observations also made it possible to demonstrate that 9 months after the injection, cells from GRPNOPC1 developed in the lesion areas.

Twelve months after the injection, the histological sections of the spinal cord of the animals treated showed *the absence of teratoma* compared to two control animal groups. The positive animals had been treated with a GRNOPC1 solution voluntarily containing 5% hESCs or containing a pure solution of hESCs. In a very small number of cases, cysts appeared but did not have any adverse effects on the animals. Concerning the acute toxicity of GRNOPC1 cells, hematological studies and analysis of the chemical constants did not lead to a significant increase in toxins being shown in relation to the control group.

In vitro studies have shown that GRNOPC1 cells are minimally recognized by the human immune system, do not aggregate with serum and are not destroyed by natural killer (NK) cells or thymic T-cells.

The transplant consists of an *injection of 2 million SCs at the site of the injury*, by a skilled surgeon assisted by a machine allowing precise control of the point of entry of the cells. Indeed, the injection of these cells must be performed with *great precision onto the exact areas of the lesions*. For this, Geron has developed a technical device specific to the surgery as well as special needles. Neurological tests are performed before and after surgery to ensure the safety of the GRNOPC1s. Subjects receiving the injection will be immunosuppressed using a low dose of tacrolimus. This immunosuppressant is gradually decreased and stopped 60 days after the intervention.

In August 2009, the *FDA suspended the authorization* granted to Geron, pending analysis of additional preclinical data. In November 2011, Geron ended this clinical trial on the grounds of a lack of financial resources [BRI 12]. Four patients underwent phase 1 of the trial, and *no improvement* in their condition was observed. However, no serious side effects were observed either. Geron's withdrawal leaves Advanced Cell Technology, another US biotechnology company based in Boston, Massachusetts, as the only company to conduct a clinical trial involving hESCs.

3.3.3.2. *Encouraging results in the fight against blindness*

Degenerative diseases of the retina are among the most common causes of blindness in the world. A total of 10 million people in the United States are believed to be affected. While most patients present an age-related macular degeneration (AMD), which causes progressive loss of vision, a minority of patients have Stargardt macular dystrophy.

This new authorization from the FDA has allowed Advanced Cell Technology (ACT) to begin a clinical trial of 12 children with *Stargardt disease*, an eye disorder associated with degeneration of the retina. This new trial involves the replacement of *cells in the retinal pigment epithelium*, altered in Stargardt disease, due to a gene mutation. Stargardt disease affects 80,000–100,000 people in Europe and the United States, making it one of the most common pathologies of childhood blindness. This hereditary form of degeneration can develop from the age of 6 and is incurable.

In November 2010, ACT launched a clinical trial in the United States, conducted using ESC derivatives in order to be able to treat this degeneration of vision. Two American patients participated in this trial, and the results proved rather promising. In September 2011, ACT announced the launch of the first clinical trial conducted in Europe from hESCs. Authorized by the Medicines and Healthcare Products Regulatory Agency (MHRA) and the Gene Therapy Advisory Committee, this trial aims to find a treatment for Stargardt disease. At the same time, rat trials resulted in complete restoration of eyesight in animals with no side effects [WAN 10]. The success of this phase I clinical trial on hESCs, involving *19 patients*, was published in the Lancet in 2015 [SCH 15], which reports the following results:

– *the absence of side effects*, cancer or rejection of the retinal cells injected;

– *improvement after 6 months in half of cases*, that is, in four out of nine patients with *AMD*;

– *improvement also in half of the cases of Stargardt disease*, compared to the untreated eye;

– some secondary problems related to the surgical procedure itself;

– the development of a cataract in four cases.

For ethical reasons, the study does not include a control group having undergone the same surgery without cell transplantation. Thus, the results are significant, and the improvements, which are minimal, affect half of the patients. As this is a phase I trial, the dosage of the injected cells remains limited. During phase II optimization, *doses will be increased* and improvements will potentially be more significant.

A Korean team of researchers published in Stem Cell Reports (2015) results on four patients with macular degeneration [SON 15]. This clinical trial was conducted using hESCs in collaboration with CHA University (Korea) and Robert Lanza's team at *Ocata Therapeutics* (USA), formerly known as ACT (Advanced Cell Technology), bought by the Japanese group *Astellas Pharma*. Identical results were obtained in the United States in 2014 on 18 patients. Indeed, these preliminary tests only proved the non-toxicity of the treatment, but not yet its effectiveness. On the contrary, the small number of patients tested does not make it possible to draw conclusions as to the reproducibility of the treatment.

3.3.3.3. *Treatment of individuals with AMD, co-developed by the Pfizer laboratory*

The first patient was operated on in September 2015 at London Moorfields Eye Hospital as part of the *London Project to Cure Blindness*, a clinical trial using hESCs in order to treat individuals with AMD. The treatment, co-developed by Pfizer, hopes to find a solution to this degenerative disease. The operation is the first in a series of 10 scheduled for this Phase I clinical trial aimed at testing the safety and efficacy of the transplantation of retinal cells, created in the laboratory from hESCs. Another clinical trial based on these same cells is underway in Israel (Hadassah University Hospital in Jerusalem). The research team is led by Profs. Benjamin Reubinoff and Eyal Banin, both of whom have already worked on the culture conditions of hESCs to differentiate them into retinal cells [IDE 09].

Macular degeneration is the leading cause of vision loss in the Western world. The variant treated by the operation, the so-called "wet" degeneration, occurs when retinal cells die as a result of the appearance of abnormal blood vessels that damage the back of the eye. To repair this, the British team first recreated part of the damaged tissue – the retinal pigment epithelium – in the laboratory using ESCs. They successfully implanted a microscopic patch containing approximately 100,000 SCs in the retina of a patient with AMD. The first results are promising: the cells seem to be developing well and "appear to be healthy" in the words of the co-leader of The London Project to Cure Blindness, Prof. Peter Coffey [RAM 13].

3.3.4. *Induced pluripotent stem cell-based clinical trials: first clinical trial on AMD in Japan*

Unlike ESCs, iPSs do not pose a fundamental ethical problem. They are created from adult cells of the patient, brought back to quasi-embryonic state by making them again express four genes, normally inactive in adult cells. They are able to transform into any cell type and potentially have the capacity to treat a large number of pathologies. To date, iPS injection is not authorized in France. Indeed, there are additional studies underway aimed at certifying the safety of these cells. It must be certain that the injection of iPSs after differentiation does not induce tumor formation at the injection site and that the IPSs are very much immunocompatible with each patient. Clinical trials have, however, begun in the United States and Japan, which remains a pioneer in iPS. It will have a specialized research center in 2019, which will be led by Prof. Shinya Yamanaka, a Nobel Prize winner for his discovery of iPSs in 2012.

In July 2013, Masayo Takahashi's research team at the Riken Center for Developmental Biology, in collaboration with the Institute for Biomedical Research and Innovation, obtained authorization from the Japanese Ministry of Health for *clinical research on AMD*. Retinal pigment epithelium cells produced from somatic cells will be transplanted to the same patient in the hope of restoring her visual abilities [JIN 11].

Three surgeons, including Yasuo Kurimoto of Kobe City Medical Center General Hospital, implanted a thin film of retinal cells (3 mm thick) in the patient's eye. The team had developed this film of retinal cells, derived from adult skin cells of this Japanese patient's arm. Two out of six planned patients were transplanted.

One year after the operation, Dr Takahashi noted a "visual improvement". For the second patient, however, the iPSs obtained from his skin cells contained mutations that were not present in the initial cells. *The team therefore decided to suspend the study*, in order to determine the origin of these mutations, to possibly change the course of the trial. The team is considering the possibility of using allogeneic iPSs from an *iPS cell bank rather than those of the patient.*

This new strategy will also significantly reduce the costs of transplanting iPSs. The first transplant on a patient with AMD in 2014 cost US $931,000. Using donor iPSs, this amount could be reduced to one-fifth.

Professor Takahashi is also behind an iPS-based clinical trial project aiming to treat Parkinson's disease. The launch of the clinical trial is delayed for safety reasons. If this therapy is approved, it will be the first in the world to use iPSs to treat a neurodegenerative disease. The research team plans to create iPSs from the patient's blood cells, evolve them into neurons and then transplant them into the patient's brain, thus counteracting the disease.

4

Stem Cells for Regenerative Medicine in Humans

> *"Why shouldn't we back up that which is most vital to our livelihood: our own stem cells at the core of every part the body?"*
>
> (André Choulika, CEO Cellectis)

Regenerative medicine aims to restore damaged tissues or organs. Based on the knowledge of cell and tissue development, it uses mechanisms where the cell is the central unit of reconstruction. It has been ten years since Shinya Yamanaka proved the existence of iPSs, or induced pluripotent SCs. From specialized adult cells, skin for example, researchers are able to return to the embryonic stage and reprogram them. The "rejuvenated" iPSs are then able to transform into any other cell of the organism. Easier to access than embryonic cells, iPSs raise fewer ethical issues. Thanks to them, regenerative medicine is gaining ground in Japan and throughout the world.

Regenerative medicine requires the development of biomaterials, indispensable supports for proliferation and cell transplantation. Thanks to advances in *tissue engineering*, the combination of SCs with *nanomaterials* makes it possible to mimic the *in vivo* environment of cells. By enabling faster cell proliferation and better differentiation, the coupling of SCs to *nanostructures* paves the way for new prospects within the field of regenerative medicine. Every tissue in the organism, whether it is bone, skin, nerve or cardiac, can be damaged as a result of disease or trauma. If they are biocompatible, the *nanobiomaterials* can be shaped to give them a structure close to the physiological tissue. This three-dimensional frame will then be colonized by the *ex vivo* SCs or the cells of the target tissue *in vivo*.

The aim of this chapter is to understand the concept of regenerative medicine and to estimate its technological feasibility. To begin with, we will look at the technologies for realizing *biomaterials in cell engineering*, and secondly, we will detail the technologies of artificial nanobiocomposites, and more specifically *nanofibers*. This technique has made it possible to carve the SC niche on a cellular scale. We will then go on to see how *bioprinting* can achieve not just a cell support but reconstruct an entire organ. These new constructions invent the regenerative medicine of today, of which some examples of marketed products will be presented at the end of the chapter.

4.1. Biomaterials in tissue engineering

Tissue engineering assembles the scaffolds, cells and biologically active molecules into a functional tissue. Biomaterials are medical devices designed to interact with living systems, providing physical structures and a support for tissue engineering. The combination of chemical and physical factors makes it possible to imitate the SC environment at *nanometer* scale. It represents one thousandth of a micron, or 10^{-9} meters, the distance between two atoms, which is the reference unit of nanotechnologies. At this *nanoscale*, the biomaterial architecture defines a topography, which has the ability to influence all the properties of the SCs from adhesion to the surface of the support, through migration, proliferation and commitment in a specialized lineage. Biomaterial manufacturing technologies in tissue engineering are oriented either towards a porous scaffold using fibers on which the SCs adhere, or towards the inclusion of SCs in the polymer, the aim being to design a functional construct, ensuring, to begin with, cell adhesion.

4.1.1. *The adhesion of cells to biomaterials*

Chemical or physical architectures share similarities with the *extracellular matrix* (ECM) and provide an optimal three-dimensional microenvironment for SC attachment. Some new nanomaterials are used as cell culture and differentiation supports because they reproduce the native ECM, a key tool for tissue engineering. Most ECM proteins form a network of fibrils from a few tens to a few hundred nanometers, or even a few microns at most. When cells interact with the ECM, anchoring points are created at the ECM by a *focal adhesion complex* associated with transmembrane proteins, the *integrins* [GIA 99]. These proteins are composed of α and β subunits, which determine the specificity of ECM anchoring [GIA 99]. Activation of integrins is regulated at nanoscale by interactions between the cells and the ECM.

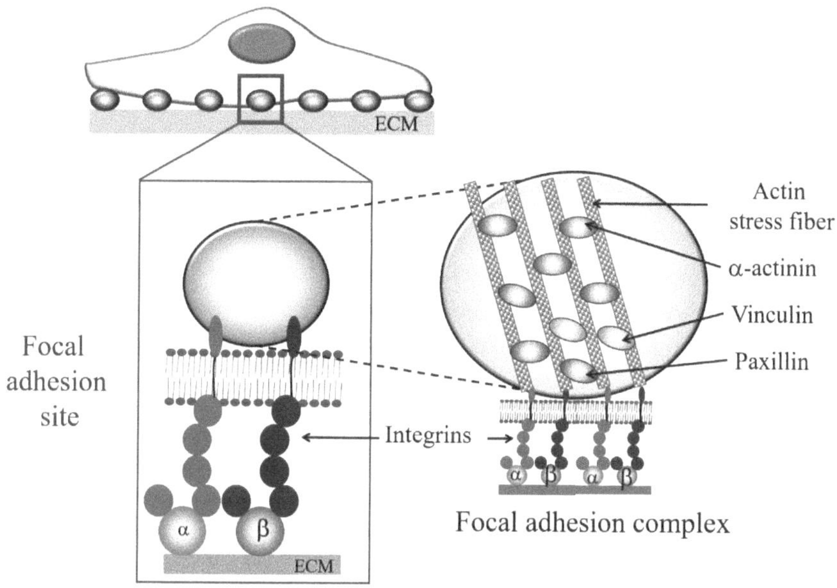

Figure 4.1. *Adhesion of SCs to the extracellular matrix thanks to the focal adhesion complex. For a color version of this figure, see www.iste.co.uk/arrighi/stemcells.zip*

4.1.2. *The extracellular matrix, a complex architecture to imitate*

The natural microenvironment of cells is the ECM, composed of fibrous proteins such as *collagen*, and non-fibrous proteins such as proteoglycans, glycosaminoglycans and soluble factors. Consistently, cells secrete, assemble and remodel the ECM. Using surface receptors, cells attach to the ECM and can activate migration, proliferation and differentiation. The composition of the ECM varies from one tissue to another, as do its structure and its biomechanical properties. The components of the ECM are key parameters in the design of *implantable biomaterials*, which generate a functional, biocompatible and biodegradable architecture. These biomaterials are designed either by chemical modification (i) of a support matrix by surface coating or the insertion of small molecules as growth factors, or by physical methods such as electrospinning (ii) and bioprinting (iii), which draw a complex architecture matrix, the composition and topography of which are mastered.

The surface *coating* plays an important role in tissue engineering because it can increase cell adhesion. This simple method makes it possible to fasten components of the ECM onto a polymer substrate, by non-specific physical adsorption. For example, the collagen IV coating allows iPSs to differentiate into functional cells of the cardiovascular and hematopoietic lineages [SCH 08]. Compared with the covalent bond, the adsorption is less strong, which may be a disadvantage of the functional implant. However, this may become an advantage for the ECM component release systems in the tissue environment after implantation. These coating techniques offer the advantage of being conducted in aqueous solution.

Electrospinning technology generates fibrous, porous matrices used in regenerative medicine. The size of the fibers and pores makes it possible to control the responses of the cells, such as migration, proliferation and differentiation. The electrospinning material consists of a syringe connected to the polymer solution. When the polymer is pumped into the syringe, a droplet is formed at the end of the needle. By applying a high voltage, the droplet takes a conical shape and as soon as the electric field is greater than the surface tension, a fine fiber develops. Electrospinning technology is straightforward and makes it possible to generate three-dimensional fibers of controlled diameter, called *3D fibers*. The major disadvantage, however, is the action of the high electric field, which can denature the functions of the proteins.

Bioprinting technology is a computer-controlled technique that enables the manufacture of a layer-by-layer support. Bioprinting consists of the addition of cells, growth factors and other functional elements in superimposed layers. The ECM proteins used in the generation of three-dimensional matrices or *3D matrices* are collagens, fibrin and gelatin methacrylate. Thanks to the deposit of material and the organization of 3D structures, it is possible to create whole organs.

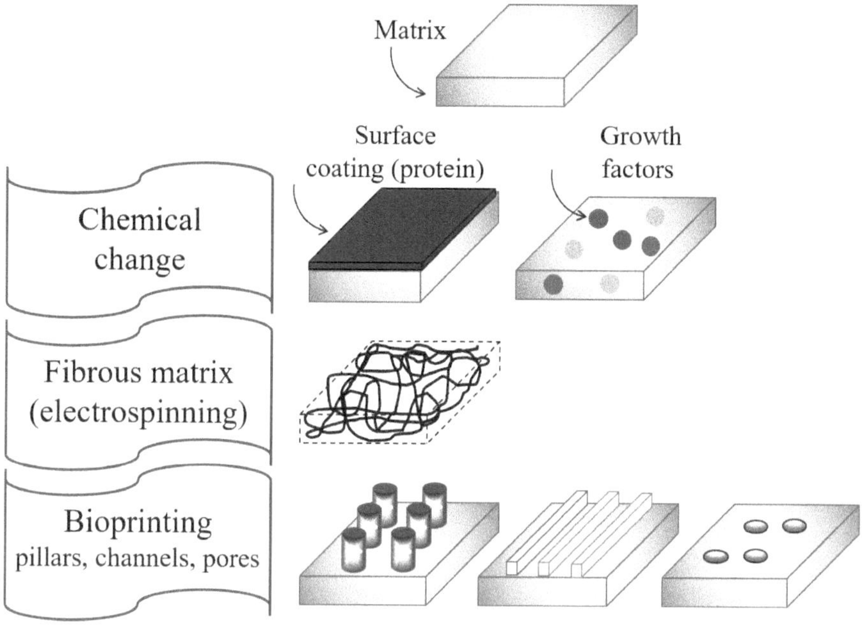

Figure 4.2. *The different types of biomaterial manufacturing technologies, imitating the microenvironment and extracellular matrix of SCs*

4.2. Nanofibers associated with stem cells organized in 3D

In the manufacture of an artificial tissue, the first step is the collection of the SCs, then the cultivation around a matrix that guides the cell growth. This matrix must ensure the proliferation and functionalization of the cells, which will be implanted into the patient. In concrete terms, the artificial environment decreases the differentiation capacities and the cells can express atypical proteins (inflammation). Nanotechnology tools make it possible to structure matrices and to control the surface and the porosity of the materials influencing the proliferation and the differentiation. The surfaces can be nanostructured by mechanical or chemical processes.

These nanotechnologies influence cell adhesion because they interact with the integrins. They also influence the capacities of self-renewal and thus maintenance of the stem nature and the hematopoietic, neurogenic, osteogenic and cardiogenic differentiation capacities.

4.2.1. *Nanotechnology and integrin adhesion*

Park and Huang's works have shown the influence of nanofiber topography in SC response. SC adherence varies according to the diameter of the fibers and their spacing at nanometer scale. It is improved in a structure of thin, close tubes, via the activation of integrin-dependent signaling pathways. A collagen protein sequence identifies a binding site with the integrin sites. This synthetic peptide confirms the activation of cell adhesion; it is of major interest in tissue engineering.

Park *et al.* [PAR 07] assembled titanium sheets polarized into TiO_2 nanotubes. These vertically oriented nanotubes have a diameter of between 15 and 100 nm. Human multipotent SCs, placed on this substrate, showed that they adhere better on thin nanotubes (< 50 nm) and that proliferation and migration were optimal with nanotubes spaced 15–30 nm apart. Cells deposited on the thin nanotubes (15 nm) present focal adhesion sites integrating β1 integrin and paxillin molecules, whereas the cells placed on 100 nm nanotubes present very little adhesion. The extracellular signal of integrin regulates a kinase signaling pathway that may be responsible for *nanotopographic effects*. These results indicate that a spacing of less than 50 nm is necessary for the grouping of integrins essential for cell adhesion and spreading on the ECM substrate. Huang *et al.* [HUA 09] tested integrin cell adhesion using nanopatterns of ordered and disordered gold particles. The results revealed that the *ordered substrate* of the nanomotif with a distance of 70 nm between ligands presented more integrin clustering. This confirms the role of the ligands necessary for the activation of integrins.

It is known that ECM proteins form a network of fibrils. The nanohelical form and periodicity of ECM proteins are important because mutations occurring in structural genes, such as procollagen-1, for example, can alter the periodicity of collagen fibrils and lead to diseases such as osteogenesis imperfecta or brittle bone disease. Another natural helical structure is that of *collagen*, which presents a unique triple helix of motif (Gly-X-Y)n. The cellular functions regulated by collagen are relayed by the interactions with the specific epitope, GFOGER (Gly-Phe-Hyp-Gly-Glu-Arg) binding to the integrin. To elucidate the adhesion mechanisms between this triple helix and the integrin, Yamazaki *et al.* synthesized a peptide mimicking the collagen triple helix [YAM 10]. They discovered that the integrin binding sequence, *GFOGER, in the synthetic peptide induces an adhesion* and spreading of the cells comparable to those obtained with native collagen. This adhesion is relayed by the integrin α2β1. This suggests the possible *application of this synthetic, triple-helix peptide in tissue engineering* and the importance of nanotopography in the design of regenerative medicine supports.

4.2.2. *Nanotopography orients the fate of stem cells*

The topographic markers vary *in vivo*, but these size and shape parameters influence cell adhesion and differentiation. However, what about pluripotency? Studies exploring nanotopography show that the stem nature is maintained in mesenchymal SCs. It is also conserved when combined with chemical factors, such as the synthetic ECM anchoring peptide. This pluripotency is also maintained on murine ESCs.

Using different nanofabrication methods, Dalby *et al.* generated nanopatterns comprising different arrangements on a polymethyl methacrylate matrix. The relief of these structures creates a controlled disorder in which hollows of 120 nm in diameter and 100 nm in depth are distributed. The space between hollows is set at 300 nm [DAL 07]. Placed on these nanotopographies, human mesenchymal SCs spread over a large area presenting several focal adhesion sites. The cytoskeleton is organized and the osteogenic markers increased compared with the cells deposited on highly ordered matrices. These works demonstrate that the use of a *nanoscale disorder stimulates the production of bone mineral matrix in vitro*, in the absence of osteogenic supplements. Efficacy is similar between topographically treated cells and those cultured under osteogenic conditions.

The topographic indications are crucial for the development of SCs; they can be combined with chemical factors. Taking this idea forward, Wang *et al.* play at the same time with *topographic and chemical modifications* during the differentiation of C2C12 myoblasts resulting from skeletal muscles [WAN 12]. The ribbed polystyrene matrix, comprising grooves that are around 900 nm wide, edges of 600 nm and a depth of 665 nm, is coupled to the *RGD peptide*. This *tripeptide of Arginine-Glycine-Aspartic Acid sequence (Arg-Gly-Asp or R-G-D), is a cell recognition and attachment site.* Proteins possessing this RGD motif, associated with integrins, constitute a major recognition system for cell adhesion. Integrin anchoring activity of the adhesion proteins can be reproduced by short synthetic peptides containing this RGD sequence. The results of the works of Wang *et al.* showed that *coupling to the RGD peptide increased both the adhesion, proliferation and differentiation* of C2C12 cells. In addition, topographic anisotropy promotes the orientation of myoblasts and the formation of aligned myotubes, as observed in muscle fibers *in vivo*. This study shows the role of chemical and topographic parameters in muscular myogenesis and the benefit of applications in muscle engineering. An adequate topography and a chemical surface provide an environment conducive to the commitment of human mesenchymal SCs to specific lineages.

4.2.2.1. *The influence on pluripotency*

Recent studies have used electrospun nanofibers creating a *three-dimensional porous architecture* based on synthetic or natural polymers, in the presence of ESCs. The first ESC studies used fibroblasts as feeder cells. The factor LIF (leukemia inhibitory factor) was then used to maintain the ESCs in undifferentiated state. The layer of feeder fibroblasts was replaced by a *biological protein matrix (Matrigel)* or a laminin coating, where the ESCs remain at the undifferentiated stage for around a hundred doublings, without losing their stem nature. To get closer to the microenvironment of SCs, Liu *et al.* generated electrospun nanofibers based on polymethylglutarimide (PMGI), which is a thermoplastic synthetic polymer stable under culture conditions [LIU 12]. The topology and density of PMGI nanofibers are key factors in maintaining self-renewal and undifferentiated state. Mouse ESCs were cultured on electrospun nanofibers of PMGI. Their proliferation is similar to that of cells in a conventional culture medium. Their *pluripotency is maintained*. The cells differentiate into three embryonic layers; this confirms that this method is feasible for the amplification of murine ESCs without feeder cells. The cellular microenvironment plays a critical role in regulating the functions of ESCs, which can only be controlled by synthetic components.

4.2.2.2. *The influence on differentiation*

The interactions between the SCs and the extracellular matrix components form part of the proliferation and differentiation control. The topographic and biochemical markers influence the commitment of the SCs to different lineages. Whether concerning the hematopoietic, neurogenic or osteogenic lineage, the properties of biomaterials make them tools for the development of tissue engineering and regenerative medicine.

4.2.2.2.1. Differentiation of hematopoietic SCs, guidance and adhesion

Chua *et al.* studied the *ex vivo* expansion of CD34+CD45+ hematopoietic SCs from human umbilical cord on electrospun nanofibers and polyethersulfone films [CHU 06]. These fibers, with a mesh of 500 nm in diameter, were grafted with hydroxyl, carboxyl or amine groups. After ten days of culture, the expansion of CD34+CD45+ hematopoietic SCs is increased in the amino nanofibers. The amplification factor obtained on these amino nanofibers is close to 200, whereas it reaches a value of 50 on a standard polystyrene support. At the same time, the amino fibers show a higher level of adhesion to the amine films and a higher percentage of progenitor hematopoietic cells. This study highlights the *potential of biomaterials in the control of proliferation and differentiation of* ex vivo *hematopoietic SCs.*

Interactions between hematopoietic SCs derived from bone marrow, and their local microenvironment also control migration, in addition to proliferation and differentiation. To mimic this local microenvironment and attract more bone marrow-derived hematopoietic SCs, Ma *et al.* generated electrospun nanofibers based on poly(dl-lactide-co-glycolide) and collagen type I, then coated with E-selectin, an adhesive biomolecule [MA 08]. In the presence of E-selectin coating, these *polarized nanofibers* increase and accelerate the hematopoietic cell capture efficiency from 23% to 67% in 30 minutes and from 30% to 70% in one hour of incubation at room temperature. This matrix presents a *biomimetic effect* capable of guiding SCs to their niche, applicable in the design of a specific unit linked to a growth niche, subsequently facilitating proliferation and differentiation of bone marrow-derived hematopoietic SCs.

4.2.2.2.2. Neurogenic differentiation

Nerve repair is a field of tissue engineering that aims to graft artificial nerves that could imitate the ECM and reinforce nerve regeneration. Biocomposite matrices based on synthetic and natural polymers provide *nerve regrowth conduits* and offer an alternative to autologous nerve grafting. By orienting the fibers in the matrix, nanotopography actively influences the morphology of nerve cells and the formation of nerve outgrowths.

The efficiency of the nano and microfibers matrices composed of poly L-lactic Acid, or PLLA, can be compared according to their aligned or disordered disposition. The perfectly aligned fibers are produced by electrospinning by adjusting the polymer concentration to the desired fiber diameter. As the PLLA matrix is intended for nerve repair, its properties are evaluated in the presence of neural SCs, whose morphology, differentiation and formation of neurite outgrowths are studied by microscopy. The works of Yang *et al.* enabled the production, by electrospinning, of a system of nanofibers and PLLA microfibers aligned on lamellae presenting diameters of 150 nm to 3 μm, in order to test the regulation of neural SCs [YAN 05]. The results indicate that the *direction of elongation of the neural SCs and that of the neurites is parallel to the direction of the aligned PLLA fibers* of the matrix. The orientation of the cells does not change according to the diameter of the fibers. However, the differentiation rate of neural SCs is greater in PLLA nanofibers than in microfibers – regardless of the fiber alignment – suggesting *the importance of nanotopography rather than microtopography*. On these bases, PLLA matrices can be used as a potential support in peripheral nerve repair and to accelerate functional recovery.

More complex biomaterials are designed using PLLA associated with collagen, as shown by the works of Kijenska *et al.*, who produced electrospun nanofibers based on poly (L-lactic acid)-co-poly(-caprolactone), known as P(LLA-CL) and collagen I/III [KIJ 12]. Different compositions are tested and different configurations of fibers, in aligned or disordered orientation. The morphology and the mechanical, physical and chemical properties of the electrospun fibers of the matrix are studied, as well as their biocompatibility, using C17.2 neural SCs. P(LLA-CL)/collagen I/collagen III-aligned nanofibers are created. They present a diameter of 250 nm and a tensile strength of 12 MPa. The cellular proliferation of C17.2 neural SCs shows a 22% increase in the case of aligned nanofibers of P(LLA-CL)/collagen I/collagen III with respect to the same nanofibers lacking collagen, that is, fibers of pure P(LLA-CL). The cell–matrix interactions and the expression of specific neurofilaments show that the *aligned nanofibers of P(LLA-CL)/collagen I/collagen III accelerate nerve regeneration.*

4.2.2.2.3. Osteogenic differentiation

Repairing the bone is the priority of bone tissue bioengineering. The development of synthetic or SC-containing biomaterials is a breakthrough in bone reconstruction. Flexible, resistant and porous, these devices must imitate the extracellular matrix, then allow the colonization of SCs and, finally, promote *osteogenic differentiation*. These biphasic composites mimic both the organic and mineral facets of the native ECM of the bone. Certain studies have investigated how variations in organic and mineral components can affect osteogenic differentiation of human mesenchymal SCs. In order to answer these questions, the Vines team created biphasic composite nanomatrices consisting of hydroxyapatite nanoparticles included in amphiphilic peptide nanofibers containing the RGDS cell adhesion motif, at different ratios [VIN 12]. The composition of these nanomatrices is validated by electron microscopy, the long-term viability and osteogenic differentiation being evaluated. According to the composition of the biomaterials, the highest ratio (66%) of hydroxyapatite nanoparticles to amphiphilic peptide containing the RGDS motif showed a better osteogenic differentiation of human mesenchymal SCs. The distribution and composition of the nanomatrix phases play a role in the osteogenic differentiation of SCs.

The matrix can also influence the *proliferation* of mesenchymal SCs. The team of Gandhimathi *et al.* [GAN 13] produced electrospun, biocompatible, polycarprolactone/poly (α, β)-DL-aspartic acid/collagen/nanohydroxyapatite (PCL/ PAA/Col/n-HA) nanofibers, coated with calcium phosphate, with a diameter of 150–250 nm. Human mesenchymal SCs were cultured on these matrices to study cell proliferation, osteogenic differentiation by alkaline phosphatase activity, mineralization and double immunofluorescent marking of the CD90 SC marker and expression of osteocalcin. This study shows that the *PCL/PAA/Col/n-HA matrix increases both the proliferation and the osteogenic differentiation* of human mesenchymal SCs.

In addition, the matrix may also play a role in the *migration* of mesenchymal SCs. Thanks to a new method of electrospinning, Rampichova *et al.* produced a structure of *elastic 3D polycaprolactone (PCL) nanofibrils* [RAM 13]. Whilst two-dimensional layer nanofibers prevent migration through the matrix, the cells infiltrate the three-dimensional matrix. The adhesion of human mesenchymal SCs to PCL nanofibers in 3D is greater than that of the 2D matrix. In parallel, the proliferation is increased as well as the viability after three weeks, as well as the expression of the osteogenic genes. Similar results were obtained *in vitro* and *in vivo* using 3D nanofibers containing hyaluronic acid (HA), polycaprolactone nanofibers and HA-chitosan nanofibers. All the *positive effects observed when using the 3D nanofiber matrices are related to the largest fiber surface available*, and hence, the importance of developing three-dimensional structures in tissue engineering and regenerative medicine.

4.2.2.2.4. Cardiogenic differentiation

Cardiac tissue engineering is a promising treatment for restoring cardiac function after a myocardial infarction. The choice of materials and cells and the architecture of the matrix play an important role. In this model, Sreerekha *et al* developed a composite by electrospinning based on fibrin [SRE 13]. Fibrin is a natural matrix for wound healing. It provides a natural environment for cellular attachment, migration and proliferation. The fibrin nanofibers have a diameter of 50 to 300 nm and those of the poly(lactide-co-glycolide) microfibers between 2 and 4 μm. These nano- and microfibers imitate the hierarchization of cardiac tissue. The results indicate that this matrix *increases the differentiation of mesenchymal SCs into cardiomyocytes*. The cardiac phenotype of the cells is confirmed by the presence of specific markers of cardiomyocytes such as cardiac calponin, sarcomeric actinin and tropomyosin.

4.3. 3D printing from stem cells

3D bioprinting is a fast technique for producing prototype implants thanks to a process of production by successive layers. It is based on several concepts in common with ordinary printing: office printers (3D printers), files to be printed (3D files), ink or *bioink composed of cells, biomaterials and bioactive substances*, and finally, paper (printing platform). In contrast to 2D printing, 3D printing is a complete process that requires the 3D construction to be designed in advance and the components of the bioink to be selected. In general, production activities are divided into three stages: modeling, bioprinting and post-bioprinting:

– The *modeling* includes computer-based design of architecture, 3D imaging and the selection of bioink and biomaterials.

– The *bioprinting* of the 3D structure is dedicated to a specific patient. The printer reads the stl file and successively deposits layers of liquid, powder or other compounds to construct the 3D model from a series of cross-sections. Bioinks contain viable SCs that will constitute a living 3D structure.

– Finally, post-bioprinting consists of *studying the functionalities*, both mechanical and biological, in order to develop mature tissues or organs for therapeutic applications in personalized medicine.

4.3.1. *3D bioprinting techniques*

3D cell bioprinting directly employs living cells in the production process. Different techniques are used to create living analogs of tissue or organ and each presents advantages and disadvantages in terms of resolution, print speed and cell viability. According to the printing mode, that is to say, according to the bioink deposit mechanism, cell bioprinting is differentiated into three categories based on extrusion, inkjet and stereolithography (laser).

Extrusion-based bioprinting (EBB) is the most promising approach because it allows the production of structures of a size required by the clinic within a very short time period. This technique produces a continuous filament of *bioink* from a pressurized syringe. This syringe is attached to a movable arm in the three x, y and z directions, which deposits the hydrogel in 3D. Several biocompatible hydrogel polymers are used, enabling cells to be deposited in high concentration. This technique uses pneumatic, mechanical and electromagnetic-force systems to guide the deposition of cells under pressure.

Inkjet bioprinting was the first approach developed from commercial printers. The printing is performed by continuous or discontinuous inkjet. In this case, the droplets are deposited one by one at precise coordinates, in order to construct a module of several droplets. A bioink cartridge contains the hydrogel and the cells that pass through the printhead controlled by an actuator. This technique depends on various energy sources according to the mechanisms of high-speed deposition (thermal, electrical, laser beam, acoustic or pneumatic). It presents the advantage of being simple with an accurate deposit.

Laser-direct writing (LDW) bioprinting has been developed from laser writing technologies. The layer absorbing the laser is stimulated by the laser pulse, which generates an air bubble, creating a pressure propelling the SCs onto the support. Unlike inkjet printing, the laser does not require a dispensing nozzle. This reduces the shear stresses during the deposit. This method presents good cell printing properties because it can use viscous bioinks and print cells at high density whilst maintaining their viability. Its advantages are the accuracy of the deposit and its high resolution.

No studies have compared SC bioprinting techniques. The shear stresses and mechanical stresses both play with the differentiation of the SCs. This would indicate that the laser-assisted technique and inkjet technique are superior to extrusion, because they are gentler processes. Extrusion presents two advantages, however: the printing of physiological cell densities and high-density biomaterials (viscosity).

The potential for differentiation and infinite proliferation of SCs are of interest to regenerative medicine. These SCs are very sensitive to the chemical, biological and mechanical parameters of the microenvironment, which makes bioprinting particularly suited for the creation of 3D microenvironments with defined mechanical and biochemical constraints.

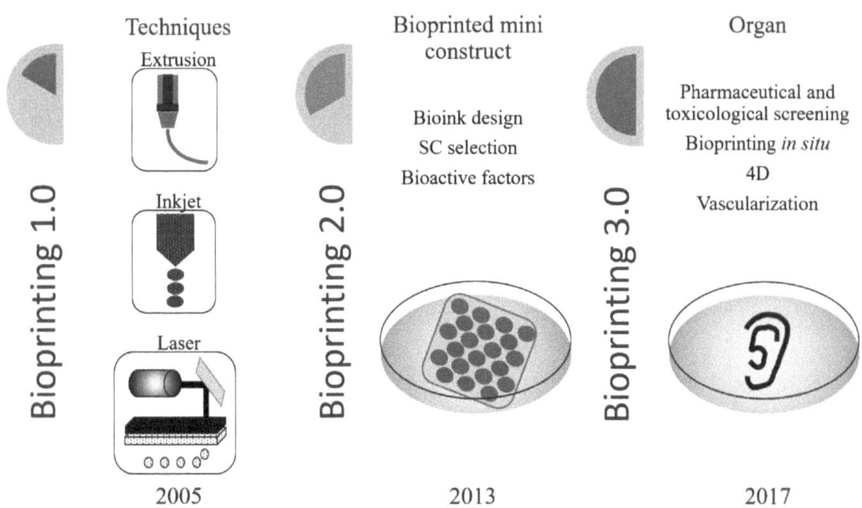

Figure 4.3. *New generation of bioprinting: This technology is based on extrusion-, inkjet- and laser-type techniques. The three essential players are the polymer (bioink), the SCs and the soluble factors that will form the matrices for 3D tissue and organ generation. For a color version of this figure, see www.iste.co.uk/arrighi/stemcells.zip*

4.3.2. 3D printing of multipotent stem cells

Multipotent SCs are present in numerous specific tissues such as bone marrow, blood vessels, the cornea and adipose tissue. They are capable of self-renewing and differentiating into specialized and limited cell types. This multilineage capacity is invested by tissue engineering in translational research, as shown by studies on the use of multipotent SCs in bioprinting in several applications in cartilage, skin and cardiac muscle regeneration.

4.3.2.1. *Regeneration of cartilage*

Bioprinting is used to print three-dimensional constructs containing living cells in a suitable biomaterial. However, limits exist as to the choice of bioink for cell encapsulation and cytocompatible polymerization mechanisms. Multipotent SCs bring great hope in translational research with 3D cartilage and bone printing specifically for each patient. This technique has also been adapted into a portable device to extrude bioinks and deposit bone or cartilage structures during the patient's operation.

Multipotent SCs were used by extrusion in the presence of silk-gelatin bioink to create a *3D tissue showing a multilineage differentiation, particularly osteogenic, chondrogenic and adipogenic potentials*. Das's team has developed biomaterials encapsulating SCs of nasal cartilage in a silk fibroin and gelatin (SF-G) bioink [DAS 15]. Gelation of the bioink is induced *in situ* by biocompatible mechanisms such as enzymatic crosslinking by microbial tyrosinase and physically via sonication. The mechanism of tyrosinases is the oxidation of the tyrosine residues of silk and/or gelatin into reactive o-quinone derivatives, which condense together. The action of sonication is to alter the hydrophobic interactions and accelerate the self-assembly of the silk fibroin macromolecules to form the crystalline sheet portion β which crosslinks the hydrogel. Sonication has no effect on the gelatin conformation. The various parameters such as the rheology, the secondary conformation of the silk-gelatin bioink and the gelation are optimized so as to increase the cell viability and multilineage differentiation of the multipotent SCs of nasal cartilage.

A new approach presented by O'Connell's works brings bioproduction into the field of surgery with the development of a *portable bioproduction tool* [OCO 16]. This "biopen" manually activates the deposition of living cells and biomaterials. A gelatin-methacrylamide/hyaluronic acid-methacrylate hydrogel (GelMa/HAMa) is printed and then crosslinked by UV during the deposition, generating 3D surgical structures. *In vitro* experiments have shown that human adipose SCs maintain their viability for one week after biopen printing in a GelMa/HAMa gel. This portable device paves the way for the use of 3D bioprinting during the surgical procedure with the possibility of directly controlling the deposition of the regeneration matrix with or without the cells.

4.3.2.2. *Reconstruction of skin layers*

Bioprinting offers the ability to create *multicellular constructs* that are difficult to obtain from other techniques. Several studies have used multipotent SCs to create skin constructs. Functional layers of human multipotent SCs, keratinocytes and fibroblasts have been successfully deposited using the laser deposit technique.

The bioprinting technique based on laser-induced forward transfer (LIFT) technology has been tested by the Koch team in the regeneration of human skin using SCs [KOC 16]. The influence of LIFT technology on SCs, survival rate, proliferation, apoptotic activity, DNA degradations and cell-surface receptor changes were evaluated over a period of several days. 98% of cells survived the transfer procedure. All cell types maintained their proliferation capacity after LIFT. Epithelial cells and mesenchymal SCs show no increase in apoptosis or fragmentation of DNA. The mesenchymal SCs retain their phenotype validated by flow cytometry (fluorescence activated cell sorting, FACS). This study shows that LIFT is a technique adapted to the *deposit accuracy of different, computer-controlled cell types that are useful in the generation of tissues* for replacement *ex vivo*.

Similarly, a laser bioprinting approach has used a hyaluronic acid–fibrinogen thrombin bioink to create a *vascular construct*. Thanks to cell–cell interactions between adult SCs and endothelial cells, this technique has made it possible to perform self-assembly of networks similar to vascular models. The works of Gruene *et al.* present a computer-assisted method for printing several cell types in a 3D matrix by laser technology [GRU 11]. By printing at given points, adipose SCs and epithelial SCs,

Gruene *et al.* show that these positions can be organized in the 3D matrix layer by layer and define several parameters: the cell–cell ratio, the quantity of cells, the combination of cell types and the distance between positions. These combinations showed interactions between adipose and epithelial SCs in the 3D matrix lacking vascular endothelial growth factor (VEGF). This demonstrates that the direct contacts between cells trigger the *development of a stable network, resembling the vascular network.*

4.3.2.3. *Cardiac muscle engineering*

The control of the deposit makes it possible to place the SCs at representative positions in the histological construction. *Cardiac muscle engineering* is emerging as a therapeutic approach capable of overcoming the limitations of cell therapy, such as cell retention and survival, and mechanically supports the ventricular wall and prevents dilation.

The studies of Gaebel *et al.* seeded human umbilical vein endothelial cells (HUVEC) and human mesenchymal SCs in an organized manner using the LIFT technique, on a polyester urethane urea cardiac patch [GAE 11]. On control patches, an equal amount of cells are disorderedly seeded without LIFT. These patches are cultured *in vitro* or transplanted *in vivo* into the infarction area of rat hearts. Cardiac performance is measured by catheterization of the left ventricle eight weeks after the infarction. Next, the hearts are perfused with fluorescent lectin in order to evaluate the functionality of the blood vessels in histological analysis. The results indicate that the models seeded with LIFT modify the growth characteristics of the HUVEC cells and mesenchymal SCs in coculture, leading to an increase in vessel formation and *functional improvement after cardiac patch transplantation.* The capillary density is increased as well as the integration of human cells within the murine vascular system.

The technique of bioprinting by extrusion was applied in regeneration of the cardiac tissue with human cardiomyocyte progenitors in an alginate hydrogel [GAE 12]. This approach made it possible to generate a tissue *in vitro* with a homogeneous distribution of cells in the matrix. The cell viability shows that 90% of the cells are viable after seven days of culture. Cardiac progenitors retain their differentiation capacity in cardiac lineage. In particular, 3D culture promotes the gene expression of early transcription factors such as Nkx2.5, Gata-4 and Mef-2c as well as the troponin T sarcomeric protein. Printed cells are also able to migrate from the alginate matrix and colonize the Matrigel layer, forming tubule-like structures. 3D cardiac cells retain their viability, cardiac lineage and migration capacity, showing the potential of 3D printing for cardiac cell administration.

4.3.3. *3D printing of pluripotent stem cells*

3D printing of multipotent SCs meets the functional needs of regenerative medicine. Unlike multipotent SCs, pluripotent SCs can differentiate into any cell type of the organism. This unlimited potential for differentiation makes pluripotent SCs attractive for regenerative medicine. However, pluripotent SCs can also differentiate into an unwanted cell type, which reduces their scope and raises the issue of ESC bioprinting.

Modeling of pluripotent SCs is not of great interest for large-scale matrix production because they have the ability to self-assemble automatically amongst themselves and change their initial configuration. 3D constructs provide a controlled environment in which the biological mechanisms of the SCs will be studied, such as the ability to self-renew and differentiate, the influence of SC colony size, distance between SC colonies and even the distance between different types of SC. By imposing only *the position of the initial deposit in a controlled environment*, 3D printing allows all possible responses of the cell and the structure.

Cellular signaling (juxtacrine, paracrine or cell–cell contact) and the type of cells involved in communication (homotypic signaling or heterotypic interactions) are the *critical parameters of the microenvironment*, particularly during tissue development. The accuracy of 3D printing allows us to deposit cells in the matrix and organize the type of cells desired, the cell gradients inside the architecture and, consequently, its cellular signaling. For example, the LDW technique was used to print SC colonies at controlled distances to explore the influence of the intercolonic space in a *juxtacrine or paracrine cell signaling mode. This spatial modeling influences the early differentiation of the ESCs* of mice, confirmed by the change in gene expression compared with random seeding [DIA 15].

4.4. Today's regenerative medicine

As of late 2016, there were 775 companies active in regenerative medicine. Organizations specialized in this sector or belonging to a multinational group develop products derived from cell therapies, tissue engineering and implantable biomaterials. Amongst them are companies that use cells as a tool for pharmaceutical screening, toxicological testing and specialized service providers in clinical research or contract production. Regenerative medicine represents the next frontier for patients as it is developing revolutionary treatments and offers the means to treat devastating diseases, some of which are not cured by conventional treatments. The rapid transposition of scientific discovery to clinic is a testament to the hope generated by these technologies. For example, the cardiac patch is a recent innovation combining tissue engineering and SCs. Other treatments have passed the experimental stage; they are authorized by the regulatory agencies in the markets of various countries.

4.4.1. *The cardiac patch*

SCs are the promise of the repair of chronically damaged organs. Comparative studies suggest that SCs committed to the cardiac lineage are more effective in improving cardiac function than those not presenting the phenotype of the cardiac cell. Menasché's team worked on cardiac progenitors derived from ESCs [MEN 15]. The first implantation of a cardiac patch was performed in 2014 within the Department of Cardiovascular Surgery at Georges Pompidou European Hospital, Paris. Prof. Menasché produced a cell patch for individuals suffering from heart failure. This patch, developed from human ESCs, was successfully implanted on a patient (authorization of the clinical trial by the ANSM in 2013), at the precise location where the infarction took place.

ESCs (line I6) are amplified and committed to the cardiac lineage following exposure to bone morphogenetic protein-2 (BMP2) and fibroblast growth factor receptor (FGF). Under these conditions, the cells express the lsl-1 cardiac transcription factor and the SSEA-1-specific embryonic antigen, used as a selection marker. The Isl-1(+) SSEA-1(+) cells are included in a fibrin matrix. In order to be able to implant these cells of interest onto the patient, it was necessary to invent a support: *a cellular patch*. According to Menasché, "the biomaterial, fibrin, was more practical, since it is already used as a surgical glue". The patch is a combination of fibrinogen and thrombin, which will polymerize in order to allow cell integration and its deposition on the infarction area. This patch is grafted onto the epicardium (the outer layer of the heart), at the area to be treated. Measuring 20 cm^2, this patch enables coverage of a relatively large area.

The implanted cells present a high degree of purity, 99%, and are SSEA-1 (+). They have lost the expression of pluripotency markers SOX-2 and NANOG, and strongly express Isl-1. No complications occurred during placement of the patch or during the postoperative period. After three months, the patient's condition had significantly improved. A new contractility could be seen on the echocardiogram in the treated area. No complications of arrhythmia, tumor formation or adverse events related to immunosuppression were detected. This first case study shows the *feasibility of generating cardiac progenitors derived from ESCs and combining it with a tissue construct*. This platform could be transposed to the generation of different lineages of progenitors derived from ESCs.

4.4.2. *Treatments authorized by the FDA*

SC transplants have accelerated in the United States. This commercial activity has prompted a series of ethical, scientific, legal, regulatory and political questions. The most important question is whether promoting autologous-, allogenic- or iPS-based SC therapies exposes clients to non-compliant interventions. These practices raise ethical questions as to the *safety and effectiveness of these interventions*, the accuracy of communication, the quality of consent and the exposure of vulnerable individuals to unjustified risk.

Turner's study [TUR 16] in the review *Cell Stem Cell* in June 2016, shows that, in the United States, 351 biotechnology companies offer *non-approved SC-based treatments*. Through direct marketing, these companies target the consumer and practice interventions in 570 clinics in the United States. The promotion of SC-based interventions promises treatment of numerous diseases. Certain clinics target niche markets; for example, some cosmetic surgery clinics communicate on facelift or breast-augmentation procedures using SCs, as well as the augmentation of sexual performance through SCs. Orthopedic and sports medicine clinics extol the virtues of SCs for joints and cartilage. Others have a broader approach and promise SC transplants for more than 30 diseases. These companies target all possible treatments: neurological and degenerative disorders, spinal cord injuries and immunological, cardiac, pulmonary, ophthalmic and urological diseases. *Most complaints reveal ethical issues arising from a lack of safeguards for safe and effective treatment.* Other complaints highlight *non-compliance with the regulations or non-compliance with the instructions of the regulatory bodies.*

As of January 17, *2017, there are 11 cell-therapy products authorized by the FDA.* Of these 11 products marketed, seven of them relate to the *hematopoietic SCs authorized to treat certain blood cancers* or immune diseases. Hematopoietic progenitors are derived from umbilical cord blood and the placenta. These cells migrate to the bone marrow where they differentiate into red blood cells, white blood cells and platelets, which join the bloodstream and restore patients' cellular deficiencies.

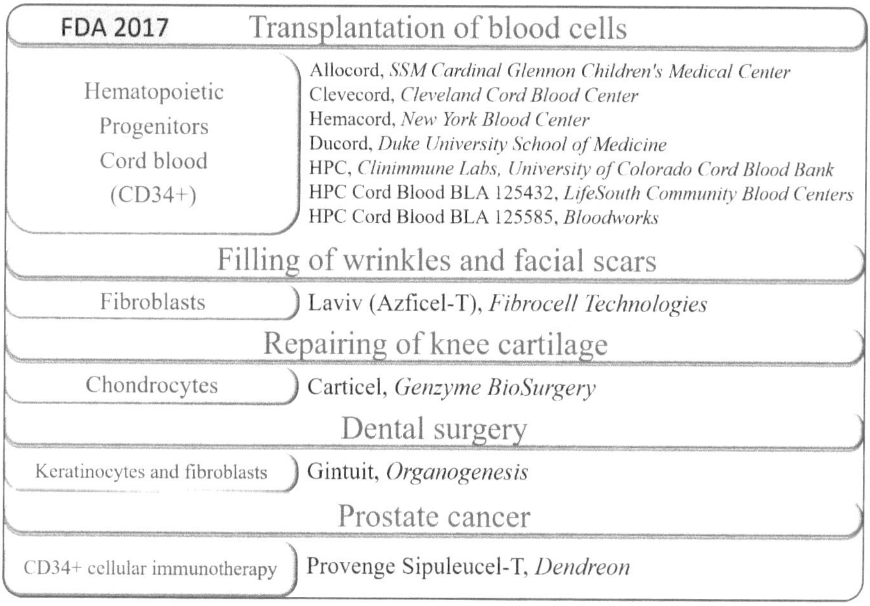

Figure 4.4. *Regenerative medicine treatments based on SCs approved by the Food and Drug Administration (FDA) in January 2017. For a color version of this figure, see www.iste.co.uk/arrighi/stemcells.zip*

4.4.3. *Authorized treatments throughout the world*

The 2016 Alliance for Regenerative Medicine report compiles key data from more than 750 international companies in cell therapy, gene therapy and tissue engineering [ALL 16]. In 2016, global funding for regenerative medicine in the United States reached $5.22 billion, whereas the figure was $1.06 billion in Europe, divided between the following three technologies:

– *Cell therapy* ($2.97 billion in the USA and $571 million in Europe).

– *Gene therapy and gene-modified cell therapy* ($2.73 billion in the USA and $353 million in Europe).

– *Tissue engineering* ($461 million in the USA and $209 million in Europe).

Cell therapy products are licensed by international regulatory agencies, as well as products marketed in the United States, approved by the Food, Drugs and Cosmetics Act and the Public Health Service Act.

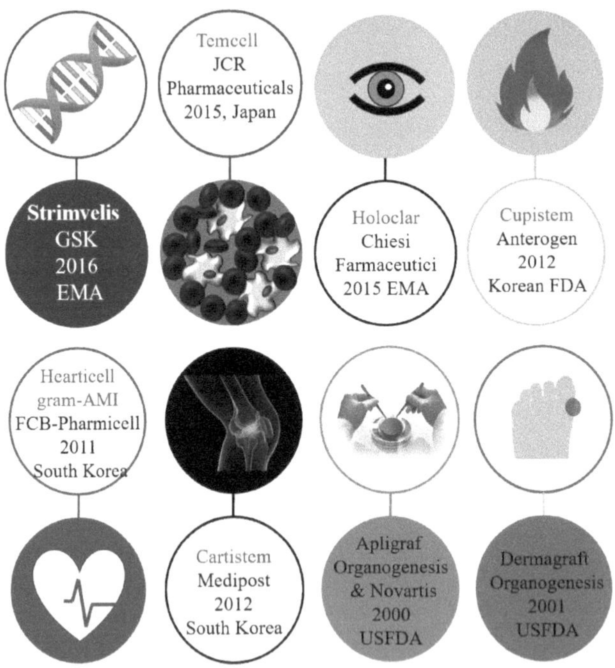

Figure 4.5. *Examples of regenerative medicine treatments based on SCs, approved worldwide as of the second quarter of 2016*

Strimvelis (GSK) was approved in May 2016 by the *European Medicines Agency (EMA)*. It is a drug used for the treatment of a severe combined immunodeficiency due to an *adenosine deaminase deficiency* (ADA-SCID). ADA-SCID is a rare hereditary disease characterized by a modification (mutation) in the gene necessary for the production of an enzyme called adenosine deaminase (ADA). This mutation causes the ADA enzyme to be absent in patients. As this enzyme is essential for maintaining healthy lymphocytes, the immune system of patients with ADA-SCID does not function correctly and without effective treatment, these patients rarely survive for more than two years. Strimvelis is used in patients with ADA-SCID who cannot be treated with a bone marrow transplant because they do not have a compatible donor amongst their family members. Strimvelis contains *cells derived from the CD34+ patient's own bone marrow that are genetically modified to contain a functional ADA gene*. Strimvelis is a type of advanced therapy drug referred to as a "gene therapy product". It is a type of drug that works by releasing genes into the organism. Given the low number of patients affected by the ADA-SCID, this disease is said to be "rare". This led to Strimvelis receiving the designation "orphan drug" (a drug used in the treatment of rare diseases) on August 26, 2005.

Temcell (JCR Pharmaceuticals Co Ltd, licensee of Mesoblast Ltd) was approved in October 2015 by the Japanese Ministry of Health, Labor and Welfare and conditionally approved in Canada and New Zealand (also called Prochymal). Temcell is an *allogeneic cell therapy product* for the treatment of severe radiation injury, chronic obstructive pulmonary disease, Crohn's disease, graft-versus-host disease (GvHD), type I diabetes and myocardial infarction. Since 2012, Canada has approved Prochymal (Remestemcel-L), indicated for graft-versus-host disease (GvHD), based on *adult mesenchymal human SCs, indicated for clinical use in children*. The Prochymal produced by the American laboratory, Osiris Therapeutics, is administered to children with a rare disease, called acute graft versus host, for which steroid treatments are ineffective. It occurs when the recipient of a bone marrow transplant is the victim of a lymphocyte attack from an incompatible donor. The symptoms are varied (abdominal pains, hair loss, digestive and pulmonary disorders) and prove fatal in 80% of children. Until then, no alternative could be offered to these patients.

Holoclar (Chiesi Farmaceutici) was approved in February 2015 by the European Commission; it consists of *ex vivo amplified autologous human corneal epithelial cells containing SCs*. Holoclar is a SC-based treatment used in the eye to replace damaged cells on the surface of the cornea. It is used in adult patients suffering from a moderate to severe *limbic SC deficiency* caused by chemical burns. Patients with this disorder do not have enough limbic SCs that usually act as a regenerative system, renewing the outer cells of the cornea when they are damaged and as they age. Holoclar is a type of *advanced therapy drug referred to as a tissue engineering product*. It consists of cells taken from the limb of the patient and then cultured in the laboratory so that they can be used to repair the damaged surface of the cornea. Given the low number of patients with limbic SC deficiency caused by eye burns, this disease is said to be "rare". This is why Holoclar has received the designation "orphan drug" (a drug used in the treatment of rare diseases).

Cupistem (Anterogen) was approved in July 2012 by the Korean Food and Drug Administration. Cupistem Injection is an autologous treatment of *mesenchymal SCs derived from adipose SCs* aimed at reducing *inflammation* and regenerating damaged tissue in the treatment of *Crohn's fistula*.

Cartistem (Medipost) was approved for marketing in January 2012 by the South Korean Ministry of Food and Drug Safety. This treatment helps to *regenerate knee cartilage*. It combines *mesenchymal SCs derived from human umbilical cord blood* and sodium hyaluronate. It is a *single-dose* therapeutic agent that enables regeneration of knee cartilage after joint trauma, degenerative arthritis or rheumatoid arthritis.

Hearticellgram-AMI (FCB-Pharmicell) was approved for marketing in July 2011 by the South Korean Ministry of Food and Drug Safety. This is an autologous intracoronal injection of mesenchymal SCs for the treatment of *acute myocardial infarction* (AMI). Mesenchymal SCs are derived from the bone marrow amplified and injected into the heart.

Dermagraft (Advanced Tissue Sciences, Organogenesis) was approved in September 2001 by the USFDA. It is a skin substitute used to *heal diabetic foot ulcers*. It is composed of *human fibroblasts* placed on a synthetic matrix, a culture support for dermal cells. By ensuring the stability of the dermis, the natural healing process is accelerated: the keratinocytes located around the wound migrate to cover the artificial dermis and produce growth factors until healing.

Apligraf (Organogenesis, Inc. & Novartis AG) was approved in June 2000 by the USFDA, also marketed under the name *Gintuit* for a dental-surgery application. Apligraf consists of a *skin equivalent bilayer*, that is, a dermal layer of human fibroblasts and an epidermal layer of human keratinocytes. The epidermal layer is composed of living allogeneic neonatal human keratinocytes, which have formed a stratum corneum. The underlying dermal layer is composed of bovine collagen fibrils and living allogenic neonatal human fibroblasts. The epidermal layer represents approximately 33% and the dermal layer approximately 67% of the total Apligraf unit. Apligraf is used for the treatment of *diabetic foot ulcers* and *venous leg ulcers* that do not heal after three weeks and six months, respectively. It stimulates healing by providing a natural barrier to protect the wound from injuries and infections.

5

Bioethics: Regulatory Framework for Human Stem Cells

"*Bioethics is a field of knowledge that mainly combines medical, biological, social, moral and legal considerations*".

(Jean-René Binet, Professor of Private Law at the University of Rennes 1)

The term "bioethics" comes from a contraction of the words "biology" and "ethics". Its inventor, American oncologist Van Rensselaer Potter [VAN 71], defines it as the combination of biology and human values in order to reconcile scientific progress with the survival of the human species. In its original conception, bioethics refers to the science of survival. Yet the concept has since come to be applied in a more restrictive and technical perspective, limited to the medical field [LEG 13a] (for a comprehensive study of the concept, see [BIN 17]).

Ethics introduces a moral framework into human action that borders on deontology, but under the moral principles respected by the individual, the legal rule that can be imposed by the community comes through. In other words, *ethics needs law in order to ensure its effectiveness, especially when the human body, and all its components, is concerned.*

The use of the human body for medical purposes of a therapeutic and experimental nature has always existed, without raising any particular legal difficulties [BIN 02]. *The progress of medical research* made it possible, over the course of the twentieth century, to reconcile therapeutic purpose and scientific purpose, such that the new uses of the human body, its elements and its products required a legal framework. The first interventions by legislators occurred intermittently: the Lafay Law of July 7, 1949 relating to corneal transplantation with the aid of eye donors; the Law of July 21, 1952 on blood, its plasma and their derivatives; the Caillavet Law of December 22, 1976 concerning the removal of organs [BIN 15a].

Yet the development of biomedical activities imposed comprehensive reflection on the need for a common framework (see *inter alia*: [LEN 91, MAT 93, FRA 06]), which led legislators to build a set of legal rules concerning the human body and the elements and products of this body, borrowing ethical values. This evolution was translated by *three laws, called "laws of bioethics"*:

– Law No. 94-548 of July 1, 1994 on the methods of processing data relating to health research.

– Law No. 94-653 of July 29, 1994 concerning "*respect for the human body*".

– Law No. 94-654 of July 29, 1994 relating to "*the donation and use of the elements and products of the human body, medically-assisted procreation and prenatal diagnosis*".

Given the evolving nature of the subject [THO 94], these laws were subject to an overall reform by Law No. 2004-800 of August 6, 2004 [GAL 04a, BEL 04, BOR 05, THO 05]. A second comprehensive reform of the legislative framework occurred with Law No. 2011-814 of July 7, 2011, this also "*relating to bioethics*" (see bibliography of the comments on these laws in [BIN 15b]; on the legislative development, see also [LEG 13b]). Lastly, the legislator proceeded with a specific amendment to the 2011 Law with Law No. 2013-715 of August 6, 2013, "*authorizing, under certain conditions, research on the embryo and embryonic stem cells*" [BIN 13, LER 13].

This legislative framework is included in the French Civil Code for the general principles governing the matter: *dignity of the person, inviolability of the human body and non-ownership of the body and its products and elements* (C.civ. arts. 16 to 16-14), and in the French Public Health Code (*Code de la santé publique*, CSP) for specific applications [BIN 15c].

The principles adopted in France within the framework of the bioethical Laws of July 29, 1994 directly inspired, at European level, the terms of the *Convention on Human Rights and Biomedicine,* known as the *Oviedo Convention*, adopted on November 19, 1996 by the Council of European Ministers. This convention develops a common framework for questions concerning the field of biomedicine and takes the principle of human dignity as an essential benchmark (regarding this principle, see [LEG 13d, MAT 09]). In particular, it emphasizes respect for every person's integrity with regard to biology and medicine, the primacy of the human being over the interests of society and science, and the prior notification and consent of the person concerned by an intervention in the field of health. *It prohibits genome modifications, the creation of human embryos for research purposes, and the human body being used as a source of profit.* After authorization of Law No. 2011-814 of July 7, 2011, the convention was ratified by France on December 31, 2011 and entered into force on April 1, 2012 [BAC 12, BIN 12, MIR 12, THI 12].

From this state of the "law of bioethics" (the concept is disputed by certain authors, such as B. Legros [LEG 13c] and J.C. Galloux [GAL 04b], but it can be understood as the set of legal rules that govern biomedical techniques), it is possible to address in a more specific manner a legal environment for stem cells that fulfills a certain number of functions.

Since these cells come from the human body, three aspects of bioethics form the subject of specific legal provisions: research, the patentability of its results and the use of these cells as innovative drugs.

5.1. The legal position relating to stem cell research

The freedom of scientific research is recognized in France as a principle with constitutional value, but it is a freedom governed by a legal framework when it concerns the human being [MAT 09, PAR 09] and *a fortiori* stem cells.

There are three main families of stem cells: embryonic stem cells, adult stem cells and induced pluripotent cells (see definitions in Chapter 1 above), each subject to its own legal provision.

5.1.1. *Embryonic stem cell research*

In the bioethical laws of 1994, stem cells are not mentioned despite their research potential recognized by the National Consultative Ethics Committee (Opinions Nos. 52 and 53 of March 11, 1997, [NAU 97]). It is Law No. 2004-800 of August 6, 2004 that establishes the methods of research specific to these cells by posing a prohibition that could be the object of derogations with the agreement and control of the French Biomedicine Agency [GAL 04b, PAU 05]. Law No. 2011-814 of July 7, 2011 extends the rules enacted in 2004 to embryonic stem cells and maintains research under a regime of prohibition, with derogations subject to strict conditions [CHE 11, VIG 11].

Law No. 2013-715 of August 6, 2013 modified the regime of research into the embryo and embryonic stems to liberalize them in order to accompany the scientific progress made in this field; it applies *the transformation from a principle of prohibition with derogations to a principle of freedom under certain conditions* [BIN 13, BIO 13, LER 13]. Whilst the principle of freedom of research on these cells is now acquired, its implementation can only occur under certain conditions, which concern research authorization and the material of this research.

5.1.1.1. *Research authorization*

An embryonic stem cell research protocol can only be authorized under *four conditions* (CSP art. L 2151-5, I). First, the relevance of basic or applied research needs to be established. This research must then form part of a medical purpose, which considerably widens its field compared with the previous framework, which required major medical progress, amounting to predicting the outcome of the research. From now on, it is necessary for the project to only be related to a medical activity. It also needs to be shown that, based on current scientific knowledge, this research cannot be carried out without the use of embryos or embryonic stem cells. The project and the conditions for implementing the protocol must still respect the ethical principles relating to research on the embryo and embryonic stem cells.

Only *institutions and organizations* pursuing a research activity and holders of embryonic stem cell conservation authorization for research purposes, as well as public and private institutions and organizations pursuing a research activity and having concluded an agreement with the previous institutions and organizations, may obtain authorization to proceed with research on embryonic stem cells (CSP art. R 2151-3, II; for a comparison with embryo research, see CSP art. R 2151-3, I).

The *request for authorization* is sent to the Director General of the Biomedicine Agency by a registered mail with request of acknowledgment of receipt, or deposited against a receipt with the latter. The application is accompanied by a file containing all of the elements necessary in order to verify that the legal conditions are fulfilled, the form and content thereof being determined by the Director General. The latter also determines the periods during which requests for authorization may be submitted. The closing date of these periods triggers the start of the decision notification period. Where documents required for examination of the application are missing, the notice of receipt of the application shall determine the period within which they must be supplied (CSP art. R 2151-6, paras. 1-3).

The *research protocol is authorized by the Biomedicine Agency* after verification of the *four conditions* required by Article L 2151-5. In addition to these verifications, the agency ensures the feasibility of the protocol and the sustainability of the organization and the research team. To this end, it takes into consideration the qualifications, diplomas, experience and scientific works of the manager and members of the team, along with the premises, materials, equipment and technical processes implemented. Lastly, it assesses the means and frameworks guaranteeing the safety, quality and traceability of embryonic stem cells (CSP art. L 2151-5, III and R 2151-2).

The Agency's decision is communicated to the ministers of health and research who may jointly request, within one month, the re-examination of the file. Where there is doubt as to the respect of the ethical principles that must govern the research or its scientific relevance, the Agency shall proceed with a re-examination of the project within 30 days. In case of confirmation of the decision, the protocol is validated. Where the protocol has been refused with regard to public health or scientific research, the Agency shall proceed with this re-examination within the same period. In case of confirmation of the decision, refusal of the protocol is granted (CSP art. L 2151-5, III and R 2151-7). The Law of August 6, 2013 thus strengthened the powers of the Biomedicine Agency by removing the health and research ministers' right to veto in favor of a request for re-examination, which can only occur for the two reasons indicated above.

Within a period of four months further to the closing date of the period during which the complete file was submitted, the Director General shall notify the applicant institution or organization of the decision to authorize or refuse the application. Failure to notify at the end of this period is implicit of a decision to refuse authorization. The decision to grant authorization is published in the Official Journal of the French Republic, mentioning the name of the individual in charge of the research (CSP art. R 2151-6, paras. 4 and 5). The authorization decision constitutes an *administrative act* subject to appeal before the administrative courts. As such, the Council of State has recently rejected on appeal the request for cancellation of authorization of a research protocol on embryonic stem cells whose purpose is the study of mechanisms and genes involved in the introduction of the differentiation of human embryonic cells into hepatoblasts (EC July 28, 2017, Jérôme Lejeune Foundation case, No. 397413).

The Biomedicine Agency also has the power to *monitor and control* research protocols. Thus, the Agency may suspend the authorization or withdraw it when the implementation of the research reveals violations of the legislative and regulatory requirements or those laid down by the authorization (CSP art. L 2151-5, III). The Director General of the Agency then pronounces the suspension for a maximum duration of three months and informs the Advisory Board. The authorization may also be withdrawn further to consultation of the Advisory Board. Prior to any decision to suspend or withdraw an authorization, the holder of the authorization is given notice to put an end to their infringements or to submit their observations within the time limit set by the Director General (CSP art. R 2151-10).

In order to carry out its control, the Agency proceeds with inspections using experts who are in no way linked to the research team (CSP art. L 2151-5, III). It is also informed directly by the research officer, who presents the Agency with an annual report on the progress of the works and a final report as soon as the research is completed. In particular, these reports contain information on the destination of the stem cells. The Director General of the Agency may, for their part, at any time request a research progress report (CSP art. R 2151-8).

Any institution or organization wishing to amend an essential element of the authorized protocol must submit a new application for authorization, which will be processed in the same way as the initial application (CSP art. R 2151-9).

Institutions or organizations conducting research on embryos or embryonic stem cells are required to keep for a period of ten years, from the end of the research, the research protocol, the documents attesting to the collection of parental consent and the final report of the research as well as the register of embryos and embryonic cells (CSP art. R 2151-12).

5.1.1.2. *The research material*

The research concerns cells collected from a preimplantation embryo between the 5th and the 7th day after *in vitro* fertilization.

Collection. Research can only be carried out *from embryos that are conceived in vitro* within the context of medically assisted procreation and are no longer intended for a parental project. In addition, research requires the prior written consent of the couple whose embryos are derived, or of the surviving member of that couple [SOR 16]. To inform the choice of these individuals, they are provided with information on possible alternatives to research, consisting of the reception of the embryos by another couple or their destruction. Consent must be confirmed following a reflection period of three months, with the exception of that involved in a preimplantation diagnosis on the embryo that has revealed an incurable genetic disease (CSP art. L 2131-4, para. 4) or an embryo that was conceived *in vitro* for medically-assisted procreation but is not to be transferred or preserved (CSP art. L 2141-3, para. 3). Consent to research is revocable without cause as long as the research has not begun (CSP art. L 2151-5, II).

Importation–exportation. Embryonic stem cells can be *imported* for research purposes and the fact that they are collected abroad must not allow circumvention of French regulations, in particular the fundamental principles set out in Articles 16 to 16-8 of the Civil Code. They can similarly be *exported* for research purposes, thus favoring the circulation of these cells in order to stimulate international research [SOR 16] (CSP art. L 2151-6). Their importation for research or export must therefore be subject to prior authorization by the Biomedicine Agency (CSP, art. L 2151-6). Only holders of the authorization to carry out research and holders of the authorization to keep embryos or embryonic stem cells, issued by the Biomedicine Agency, may obtain an import or export authorization for research purposes (CSP art. R 2151-14). The Director General of the Agency authorizes the import and export of embryonic stem cells for research purposes after consulting the Supervisory Board; the authorization is valid for two years (CSP art. R 2151-15). The operation of importing or exporting embryonic stem cells is subject to the affixing to the outer packaging of a certain amount of information (see CSP art. R 2151-16). Any incident occurring during the transport of these cells must be declared to the Agency by the holder of the import or export authorization. Where the incident is likely to affect health safety, the Director General of the Agency informs the National Agency for the Safety of Medicines and Health Products and the Minister for Health (CSP art. R 2151- 17).

Any organization importing or exporting embryonic stem cells must prove that it has obtained, in accordance with the requirements of the Civil Code, the prior consent of the couple that procreated the embryo, which was conceived within the framework of medically-assisted procreation, and that it is no longer intended for a parental project, without any payment in any form being allocated to the couple (CSP art. R 2151 - 13).

The preservation of embryos and embryonic stem cells for research purposes requires an authorization issued by the Biomedicine Agency. Its delivery is subject to compliance with the general principles relating to the donation and use of the elements and products of the human body (CSP art. L 1211-1 to L 1211-9), compliance with the rules relating to the safety of individuals on-site and the protection of the environment as well as those concerning health safety (CSP art. L 2151-7, paras. 1 and 2).

The application for authorization to preserve embryos and embryonic stem cells follows the same rules as those set for the application for authorization of a research protocol (CSP art. R 2151-20, which refers to R 2151-6).

The Director General of the Agency for Biomedicine authorizes the preservation of embryos and embryonic stem cells, after consulting the Advisory Board, for a fixed term of not more than five years. The authorization mentions the name of the person responsible for preservation. It is renewable under the same conditions (CSP art. R 2151-19, para. 1).

In order to investigate the authorization, the Agency shall evaluate the conditions for the implementation of conservation, in particular by verifying that the supply and conservation of the embryos and embryonic stem cells provide sufficient guarantees in order to ensure compliance with the provisions relating to the substantive conditions of the research authorizations (CSP, art. L 2151-5), the rules relating to the safety of persons engaged in a professional activity on-site and the provisions applicable to the protection of the environment and the health safety rules. Moreover, the Agency ensures the competence of the team in charge of preservation by taking into consideration the qualifications, diplomas, experience and scientific work of the members of the team. It also takes into account the premises, materials, equipment, and processes and techniques implemented by the applicant. Finally, the Agency evaluates the means and devices implemented in order to guarantee the safety and traceability of the embryos and embryonic stem cells (CSP art. R 2151-19, paras. 2-4).

When the applicant organization carries out preparatory, distribution or transfer activities simultaneously on the same site, the Director General verifies that provision has been made for the procedures guaranteeing against any risk of contamination (CSP art. R 2151-19, last para.).

In the event of non-compliance with the conditions of issue, the Agency may suspend or withdraw the conservation authorization (CSP, art. L 2151-7, para. 3). The Director General of the Agency then organizes the transfer of these embryos or cells to another organization authorized to preserve them (CSP art. R 2151-21, para. 1). In the event of an incident likely to affect health safety, the Director General immediately informs the Director General of the Agency for the Safety of Medicines and Health Products (CSP art. R 2151-21, para. 2).

Moreover, the National Agency for the Safety of Medicines and Health Products is informed of any preservation activities carried out on the same site as the activities authorized by it (CSP, art. L 2151-7, para. 4).

The monitoring of the conservation authorization is subject to the same rules as the research authorization, in particular the keeping of a register of embryos and embryonic stem cells and their coding by the Agency for Biomedicine (CSP art. R 2151-20, which refers to Articles R 2151-8 to R 2151-12).

Any organization preserving embryos or embryonic stem cells for research purposes must be able to justify the collection of consents given by embryo procreators. When the embryonic stem cells have been imported, the organization must be able to prove that they have been obtained, in accordance with the requirements of the Civil Code, with the prior consent of the couple that procreated the embryo conceived within the framework of medically-assisted procreation, and that this embryo is no longer intended for a parental project, this without any payment in any form being allocated to the couple (CSP art. R 2151-18).

The organizations responsible for the preservation of embryos and embryonic cells for research purposes are required to keep for a period of ten years the preservation protocol, the documents attesting to the collection of parental consent and the final report and the register of the holding of the embryos or embryonic stem cells (CSP art. R 2151-20, which refers to R 2151-12).

The transfer of embryos or embryonic stem cells can only be made for the benefit of an organization holding the necessary authorization to carry out research on these elements. The Biomedicine Agency must be informed in advance (CSP, art. L 2151-7, para. 5).

5.1.2. *Adult stem cell research*

Adult stem cells are cells from specific tissues (see definition in Chapter 1) such that different types of cells exist: nerve stem cells, cardiac stem cells, adipose tissue stem cells, muscle stem cells, skin stem cells, liver stem cells, umbilical cord blood and placental stem cells, bone marrow stem cells, mesenchymal stem cells and hematopoietic stem cells. Since these stem cells come not from the embryo but from the individual who has been born, that is, once their organization is fully developed, *research on these adult cells does not conflict with the principle of respect for human dignity and is placed under the principle of freedom of research*.

Nevertheless, the collection of cells, other than hematopoietic cells, from a living person for scientific purposes may only take place on the condition that the donor, duly informed of the purpose of the collection, has given his consent in writing. This consent is revocable at any time and without any formality (CSP art. L 1241-1, paras. 1 and 2).

As regards the collection, always for scientific purposes, of hematopoietic cells of cord blood and placental blood, as well as cells of the cord and placenta, it can only take place under the conditions of being an anonymous and free donation and that during pregnancy, the woman gives her consent in writing after having received information on the purpose of the usage. The consent remains revocable without any formality as long as the collection has not taken place (CSP art. L 1241-1, para. 4).

Moreover, the French Public Health Code mentions adult stem cells for inclusion in a research comparison for the evaluation of the state of progress of embryonic stem cell research (CSP art. L 1418-1-1, II). For the rest, adult stem cells fall under the freedom of research.

5.1.3. *Induced pluripotent stem cell research*

Induced pluripotent stem cells have characteristics close to embryonic stem cells, but they are obtained by genetic reprogramming of differentiated adult cells (see Chapter 1). In the annual report of the Biomedicine Agency, addressed to French Parliament, induced pluripotent stem cells, such as adult stem cells, serve as a basis for comparison to assess the state of progress of embryonic stem cell research (CSP art. L 1418-1-1, II). Apart from this function, induced pluripotent stem cells also fall under the freedom of research.

5.2. The patentability of stem cell research

Research needs its results to be protected in order to prevent the invention it has allowed from escaping its author through scientific use and/or uncontrolled commercial exploitation. The invention patent confers a legal certainty by attributing a monopoly, the prerogatives of which are limited in time so as not to hinder technical or scientific progress. In the context of biomedical research, but not only so, its funding is heavily dependent on contracts concluded with the pharmaceutical industry, which in turn implies protection by patent law in the form of an assignment or license to operate.

However, "*any inventions whose commercial exploitation would be contrary to the dignity of a human being are not patentable*" (CPI art. L 611-17; [CHE 06, GAL 04c, GAL 04d, LAM 05]. Yet in actual fact, research on human stem cells comes under this restriction since they make up the embryo. Consequently, "*The human body, at the different stages of its constitution and development, as well as the simple discovery of one of its elements, including the total or partial sequence of a gene, cannot constitute patentable inventions. Only an invention constituting the technical application of a function of an element of the human body can be protected by patent. This protection covers the element of the human body only to the extent necessary for the realization and exploitation of this particular application*" (CPI art. L 611-18, paras. 1 and 2). In particular, the article specifies that "*usage of human embryos for industrial or commercial purposes*" (para. 3, c) is not patentable.

It is therefore appropriate to examine the application of this legal framework to different human stem cells.

5.2.1. *Patentability and embryonic stem cells*

Embryonic stem cells are undifferentiated cells that make up the embryo between fertilization and the moment when cell division reaches 64 cells. These cells, because of their existence in the very first moments of human existence, are of major importance to genetics. In addition, these cells possess the characteristic of being pluripotent, thus allowing a therapeutic usage. They are also the origin of differentiated cell lines with specific characteristics of specific tissues, muscles, bone marrow, skin, liver, etc. This shows the benefit of these cells for scientific research and for possible industrial and commercial exploitations.

Embryonic stem cell research is already subject to considerable legal constraints and, furthermore, it cannot, in principle, be patented for industrial or commercial usage (CPI art. L.611-18, paras. 1 and 3, c). Two questions thus arise: first, does scientific research constitute a use for industrial and commercial purposes and, second, what is the extent of the exclusion from patentability?

5.2.1.1. *Research and industrial or commercial purpose*

An invention, in order to be patentable, must particularly and necessarily be subject to an industrial application (CPI art. L 611-15). Admittedly, it is possible for research not to form part of this purpose but, in any event, the rights granted to the patentee inevitably confer them a monopoly on the exploitation or commercialization of the results. Consequently, and logically, it is not possible to conceive of the attribution of a monopoly of exploitation to inventions for which all industrial or commercial usage is prohibited (CPI art. L 611-18, para. 3, c; [MAR 12]).

The Court of Justice of the European Union has had the opportunity to comment, *inter alia*, on this subject in a leading judgment of October 18, 2011 (*Oliver Brüstle v. Greenpeace eV*, C-34/10, [MAR 12, HAU 12, GAL 12a, GAL 12b, MAR 11, HEN 12, MAL 12, BOU 12, CHE 12]). One of the questions submitted to it was whether the exclusion of the patentability of the use of human embryos for industrial or commercial purposes also concerned use for scientific research purposes, since in this case the patent had been filed with a view to scientific research. On the basis of Article 6 (2) of European Directive 98/44/EC of July 6, 1998, the Court decided that "*the exclusion from patentability of the use of human embryos for industrial or commercial purposes [...] also relates to the use for scientific research purposes*". The silence of the legislator, both European and French, can thus not be interpreted to mean the patentability of this type of research [MAR 12] which, beyond the specific features of research on embryos and embryonic stem cells, is in accordance with patent law. A similar decision was already taken by the European Patent Office in a judgment by its Grand Board of Appeal dated November 25, 2008 (Dec. G 2:06, Wisconsin Alumni Research Foundation, [GAL 08, GAL 09].

In addition, the exclusion of research with a scientific purpose from patentability is confirmed and reinforced *a contrario* by the possibility of filing a patent when the invention constitutes the technical application of a function of an element of the human body. Nevertheless, patent protection only covers the element of the human body to the extent necessary for the realization and exploitation of that particular application, that is, for the purposes of diagnosis or therapy (CPI art. L 611-18, para. 2).

5.2.1.2. *The scope of exclusion from patentability*

The Court of Justice of the European Union, in its decision of October 18, 2011, also ruled on the scope of the exclusion from patentability. Since the rights granted by the patent relate only to the elements claimed in the patent application, the question was whether the exclusion should include the assumption that the patent application does not claim the use of human embryos for its establishment without mentioning the use of embryos, which has nevertheless been carried out to achieve the invention. The lack of mention of such use could limit the scope of exclusion to the claims alone.

The judgment delivered by the Court is unequivocal. This absence in the patent claim does not change the scope of the exclusion: it is general. As a result, any invention that requires the destruction of human embryos or their use as a research medium is excluded from patentability. Not limiting the patentability exclusion field to the claims alone enables this exclusion to be extended to any intervention that would use human embryos even indirectly, whether or not the usage is claimed [MAR 12].

5.2.2. *Patentability and adult stem cells*

Adult stem cells are derived from different specific tissues of the human body (see definition above) and offer a great potential for research and therapy through this diversity. The *patentability* of the research results obtained on these cells *depends on the link* that they have *with the human body* and the dignity of the human being (see CPI art. L 611-17 and 611-18). The removal of these cells from the tissues *of a developed human body* does not destroy life of a human nature, unlike what happens with embryonic stem cells.

As a consequence, any research into adult stem cells may see its results *subject to a patent,* where it fulfills the conditions for issuance required by a common patent law, namely: *a new invention resulting from an inventive step and likely to be applied industrially* (CPI. art. L 611-10, 1).

5.2.3. *Patentability and induced pluripotent stem cells*

Although induced pluripotent stem cells have characteristics close to embryonic stem cells, they are nevertheless obtained by genetic reprogramming of differentiated adult cells. *Owing to this reprogramming and their origin, they no longer have a direct link with the embryo and with the principles of protection of the human being.* According to this logic, these cells are subject to the principle of freedom of research. Consequently, the results of the research of which they form the subject are eligible for the granting of a patent under the conditions of ordinary law (see CPI art. L 611-10, 1).

5.3. Cell- and tissue-based therapy products

The products of cell- and tissue-based therapies offer a huge field for the development of therapeutic innovation in France, Europe and throughout the world. Insofar as these products have curative or preventive properties with respect to diseases or make it possible to act on physiological functions by a pharmacological, immunological or metabolic action, European Directive 2001-83 of November 6, 2001 regards them as biological drugs (see Appendix I of the Directive). As drugs, their production and free circulation must be ensured by regulation that combines the rules of the market in the biotechnology sector and those that ensure the dignity of the human being.

Such drugs are subject to a specific regime stemming from European Regulation 1394-2007 of November 13, 2007, which confirms the innovative nature of cell and genetic therapies and establishes a legal definition for tissue engineering products and combined advanced-therapy medicinal products [PEI 08]. The regulation brings together the products of genetic therapy, cell therapy and tissue engineering under the status of "*advanced-therapy medicinal products*". As such, it assigns four categories of advanced therapies:

– genetic therapy drugs;

– cell therapy drugs;

– products derived from cell engineering;

– combined advanced-therapy medicinal products.

However, the European regulation applies only to advanced-therapy medicinal products prepared industrially or manufactured according to a method involving an industrial process aimed at obtaining a standardized product. As a result, *advanced-therapy medicinal products prepared on a non-routine basis are excluded from the scope of the regulation and are subject to national law*.

In French domestic law, advanced therapies came to be covered by regulation on an *ad-hoc* basis. The first advance came as a result of the Law of May 28, 1996, which established a classification of the cells, notably introducing the category of biological products. Thereafter, Law No. 2004-800 of August 6, 2004, relating to bioethics instituted three categories of products: gene therapy preparations, xenogeneic cell therapy preparations and allogeneic or autologous cell therapy preparations. Subsequently, Law No. 2011-302 of March 22, 2011 took into account the 2007 European regulation to incorporate into domestic law advanced-therapy medicinal products prepared on a non-routine basis.

As a result, *innovative therapies fall into three broad categories of products*:

– advanced-therapy medicinal products (ATMPs) consisting of genetic therapy drugs, somatic cell therapy drugs, tissue and cell engineering drugs and combined advanced-therapy medicinal products;

– advanced-therapy medicinal products prepared on a non-routine basis (ATMP-pnrb), the manufacture of which is authorized by the Member State of the European Union and intended for a given patient;

– cell therapy preparations (CTP).

5.3.1. *Classification of cell- and tissue-based products*

Each type of product is subject to a specific regulatory framework. Advanced-therapy medicinal products are subject to European law (Regulation 1394-2007 and Directive No. 2001-83) for their marketing and post-authorization follow-up procedures, whereas clinical trials fall under national law. Advanced-therapy medicinal products prepared on a non-routine basis incorporate a national regulatory framework, which nevertheless must be equivalent to the European rules on quality and safety. As regards cell- or tissue-based therapy preparations, they are not considered as advanced-therapy medicinal products because their cellular support is not substantially modified in nature and function; they also come under national law.

The existence of three statutes requires that the *qualification of the products be determined very early* in order to follow a development in accordance with the corresponding regulatory framework. This is all the more true as the characteristics of the product can evolve during the process, thus moving it from one status to another.

The first step consists of determining whether the product constitutes an advanced-therapy medicinal product (ATMP) or a cell-based therapy preparation. Two criteria make it possible to determine whether the product is an ATMP.

The *first criterion* consists of establishing whether the manufacturing process involves *substantial changes in the cells or tissues*. The substantial change is not defined or even described, the 2007 European regulation merely listing those manipulations that are not considered to be substantial (for example, cutting out, sterilization, freezing, etc.; see Appendix 1 of the Regulation). By inference, we can consider as substantial a modification of the "initial biological properties of the cells or tissues" (see ANSM Guide 2012). That said, a succession of non-substantial changes may well lead to a process that achieves the biological properties of the cells to ultimately result in a substantial modification.

The *second criterion* lies in the *essential function of cells or tissues*. The product is an ATMP or an ATMP-pnrb when the cells or tissues are not intended to be used for the same essential function(s) in the recipient and the donor. Conversely, a product whose manufacturing process does not involve substantial modifications and whose destination is the same in the donor and the recipient must be qualified as a cell- or tissue-based preparation. If, however, one of these two conditions is not verified, the product must be considered as an ATMP or ATMP-pnrb.

Again, attention should be paid to the *evolutionary nature of the status* as the modification of the manufacturing process enabling the new biological characteristics, physiological functions or structural properties to be obtained can become substantial and change the status of the product. Likewise, a product initially covered by cell therapy preparation can come to form part of the ATMPs if the main function sought becomes different.

Moreover, during each modification of the manufacturing process and each development, it is important to determine the status of the product.

The second approach refines the previous one by differentiating an ATMP-pnrb from an ATMP. The difference can easily be observed by examining the characteristics and the destination of an ATMP that is prepared *on a non-routine basis for the attention of a particular patient*. No regulatory distinction is made between a product of autologous and non-autologous origin, but it may be considered *that an autologous product naturally better meets the definition of an ATMP-pnrb* (see ANSM Guide 2012).

We can perceive the extent to which determining the status of the product is difficult. It is therefore recommended, in case of doubt as to its identification, especially between that of PCT and of ATMP, to seek the opinion of the French National Agency for the Safety of Medicines and Health Products (ANSM) which can, if necessary, direct the request to the

European Medicines Agency (EMA) and its Committee for Advanced Therapies. Classification must therefore be undertaken as soon as possible in order to allow a development process in line with the applicable regulatory framework. It is best to undertake this step prior to implementation of the studies accompanying the marketing application.

5.3.2. *Advanced-therapy medicinal products*

Amongst the four ATMPs defined by the European texts, two are prepared from stem cells: somatic cell therapy drugs and drugs derived from cell and tissue engineering.

A somatic cell therapy drug is what was previously known as a biological drug. It is determined by two sets of characteristics. First, the drug must contain or consist of cells or tissues that have been substantially manipulated to alter their biological features, of physiological functions or of structural properties or cells or tissues that are not intended to be used for the same essential function(s) in the recipient and the donor. Second, the drug must be presented as having properties enabling the treatment, prevention or diagnosis of a disease through the metabolic, immunological or pharmacological action of its cells or tissues, or used in a person or administered to a person in such a perspective. These two sets of conditions are inclusive (see definition in Appendix 1, Part IV of European Directive 2001-83 of 6 November 2001).

A drug derived from cell or tissue engineering is a "*product that contains cells or tissues derived from, or constituted by, cell or tissue engineering and that is presented as possessing properties that enable it to regenerate, repair or replace a human tissue, or is used in human beings or administered to human beings for that purpose*". Those considered as derived from cell engineering are: "*cells or tissues that meet at least one of the following conditions:*

– The cells or tissues have been subjected to a substantial manipulation, so as to obtain biological characteristics, physiological functions or structural properties useful for the sought regeneration, repair or replacement. The manipulations listed in Appendix I, in particular, are not considered to be substantial manipulations.

– The cells or tissues are not intended to be used for the same essential function(s) in the recipient and the donor" (Article 2 of European Regulation 1394-2007 of November 13, 2007).

The substantial nature of the manipulation required in the definitions of these two drugs has not been defined and the European regulation merely indicates a series of manipulations that are not essential (see Appendix I of the regulation of 2007-1394); it can thus be considered that a manipulation that brings a modification of the biological properties or of the initial function is substantial. The substantial nature may also result from a succession of modifications, each being non-substantial.

These two types of drugs are subject to European regulations governing ATMPs with regard to marketing, manufacturing, development and clinical trials; they fall under national law.

5.3.2.1. *Clinical trials*

A clinical trial consists of "*a clinical study fulfilling one of the following conditions:*

a) The assignment of the participant to a particular therapeutic strategy is determined in advance and does not fall under the normal clinical practice of the Member State concerned.

b) The decision to prescribe the experimental drugs is made at the same time as the decision to include the participant in the clinical trial, where normal clinical practice, and diagnostic or surveillance procedures apply to participants".

The clinical study, meanwhile, is defined as "*any investigation relating to the intended person:*

a) To identify or verify the clinical, pharmacological or other pharmacodynamic effects of one or more drugs;

b) To identify any adverse effects of one or more drugs; or

c) To study the absorption, distribution, metabolism and excretion of one or more drugs" (Article 2 of European Regulation 536- 2014 of November 16, 2014).

These trials are essential for the successful completion of the manufacture and marketing of the drug.

Clinical drug trials are governed by Regulation No. 536-2014 of April 16, 2014, which requires a prior authorization (CSP art. L 1124-1, I, para.1; art. 3 of the Regulation), so that they are conducted with respect to *the rights, safety, dignity and well-being of participants, taking precedence over all other interests and with the purpose of producing reliable data* (art. 2 of the Regulation). Prior authorization implies a scientific and ethical examination of the project (art. 4 of the Regulation); the application is filed with a Member State (art. 5 of the Regulation), which will draw up an evaluation report made up of two parts: Part I for the scientific examination and Part II for the ethical examination (arts. 6 and 7 of the Regulation). The ANSM is the French authority responsible for the evaluation of the first part of the report, and the Independent Ethics Committees are in charge of evaluating the second part. It is the ANSM, however, that notifies the sponsor of the single decision relating to the clinical trial consisting of an authorization, a conditional authorization or a rejection (CSP art. L 1124-1, I; art. 8 of the regulation). In the event that an authorization is refused, the proponent may refer a request for reconsideration to the ANSM for Part I of the report and the French Health Minister for Part II (CSP art. L 1124-1, II). Decisions concerning the prior authorization of clinical trials are administrative decisions that, as such, can be appealed before the administrative courts.

Clinical trials on ATMPs, including somatic cell therapy cell drugs and products derived from cell engineering, can only be performed in healthcare facilities or blood transfusion facilities, or in the medical or blood transfusion services of the armed forces (CSP art. L 1124-1, III) and in accordance with good clinical practice, which is defined as "*a set of ethical and scientifically precise quality requirements for the design, conduct, execution, monitoring, auditing, recording, analysis and reporting of clinical trials ensuring the protection of participants' rights, safety and well-being, as well as the reliability and robustness of data obtained during the clinical trial*" (art. 2 of the Regulation).

5.3.2.2. *Manufacture*

Cell therapy drugs and drugs derived from cellular or tissue engineering must be manufactured in accordance with the rules of *good manufacturing practice for medicinal products for human use* as laid down in European Directive 2003-94 of October 8, 2003.

Their manufacture can only be carried out in France in pharmaceutical establishments (CSP art. L 5124-1) or in pharmaceutical establishments created within a non-profit organization or public institutions other than healthcare facilities (CSP art. 5124-9-1), which are subject to authorization from the ANSM.

5.3.2.3. *Marketing*

The marketing of cell therapy drugs and drugs derived from cell or tissue engineering requires, like any other drug, an authorization (MA). Since advanced-therapy medicinal products are included in the list of drugs requiring authorization by the European Union (see art. 4 and the Appendix to Regulation 726-2004, point 1a), the application for authorization forms the subject of a *centralized procedure submitted to the European Medicines Agency (EMA)* for opinion. The Agency's Committee for Medicinal Products for Human Use consults, for scientific evaluation, the Committee for Advanced Therapies (created by European Regulation 1394-2007 of November 13, 2007, art. 20), which prepares a draft opinion endeavoring to arrive at a scientific consensus. If such a consensus is not reached, the Committee for Advanced Therapies adopts the position of the majority of its members and mentions the diverging opinions in the draft opinion. Where the scientific opinion of the Committee for Medicinal Products for Human Use is not in accordance with the draft opinion of the Committee for Advanced Therapies, a detailed explanation of the reasons for the differences is attached as an appendix to the opinion (art. 8 of Regulation 1394-2007). The opinion of the Committee for Medicinal Products for Human Use must be delivered within 210 days of the request (art. 6 (3) of Regulation 726-2004).

The marketing authorization is issued by the European Commission (arts. 3 and 4 of Regulation 726-2004). It remains valid for five years and can benefit from a renewal, after reassessment of the benefit–risk ratio, for an unlimited period (art. 14 of Regulation 726-2004). If marketing is not effective within three years of the issue of the authorization, the authorization ceases to be valid. Similarly, when a drug authorized and marketed is no longer found on the Union market for three consecutive years, the authorization issued for that drug ceases to be valid (art. 14 of Regulation 726-2004).

5.3.3. *Advanced-therapy medicinal products prepared on a non-routine basis*

The four types of ATMP drugs, including somatic cell therapy drugs and drugs derived from cell or tissue engineering, are found under the status of ATMP-pnrb. *These drugs must be prepared on a non-routine basis, that is to say intended for a certain patient* (see CSP art. L 5121, 17). They present the statutory characteristic of following a national regulatory framework that must be equivalent to the European rules (see art. 28 of Regulation 1394-2004, which adds a point 7, concerning ATMP-pnrb, to art. 3 of Directive 2001-83 listing the drugs to which it does not apply).

5.3.3.1. *Clinical trials*

Clinical trials of ATMP-pnrb in France require prior authorization issued by the ANSM, which conducts a scientific evaluation (CSP art. L 1124-1, I). They must also obtain the favorable opinion of a committee for the protection of individuals, which intervenes for a social and ethical evaluation (CSP art. L 1123-6). They can only be carried out in healthcare or blood transfusion facilities, in the medical services of the armed forces or in the Armed Forces Blood Transfusion Center (CTSA, *Centre de transfusion sanguine des armées*) (CSP art. L 1124-1, III).

5.3.3.2. *Activity authorization*

The *manufacture* and, more generally speaking, the preparation, the distribution, the transfer, the import and the export of the ATMP-pnrb within the framework of research can be assured only by the institutions and organizations authorized by the French Agency for the Safety of Health Products after consultation of the Biomedicine Agency (CSP art. L 4211-9-1, para. 1). These include pharmaceutical establishments, healthcare facilities, the French blood agency (*Etablissement français du sang*) and the Armed Forces Blood Transfusion Center as well as public institutions of a scientific and technological nature, scientific cooperation foundations, utility foundations and associations when these organizations are concerned with health or biomedical research (CSP art. R 4211-32, I). This authorization is issued for five years, renewable; it is renewable and can be modified, suspended or withdrawn (CSP art. L 4211-9-1, para. 2). The authorization procedure (see CSP art. R 4211-32 *et seq.*) presents certain specific elements such as, notably, an evaluation period of 90 days as of receipt of the application or, at the end of this period, a lack of decision being deemed equivalent to a rejection of the application (CSP art. R 4211-36).

5.3.3.3. *Product authorization*

ATMP-pnrb drugs require the obtaining of *authorization* from ANSM, the French equivalent of a marketing authorization (CSP art. L 4211-9-1). The request must include a certain amount of information to enable decision-making, in particular the number of patients concerned by the drug, the data relating to the tissues and cells taken, the institutions, the pharmacology system and the health facilities in which the drug may be administered (CSP art. L 5121-210, I). The applicant must be notified of the decision within 120 days of receipt of the file (compared with 210 days for other drugs: CSP art. R 5121-35). The absence of a decision as of the expiry of the time-limit is tantamount to the rejection of the application, contrary to the rule of common law, according to which silence is tantamount to acceptance (CSP art. R 5121-212). The Biomedicine Agency is informed of the decision taken by the ANSM (CSP art. L 5121-212).

The pharmacovigilance exerted on ATMP-pnrb drugs includes a follow-up to ensure a traceability between the various steps of the process, ranging from the collection of cells to administration of the drug (CSP art. R 5121-201-4). Institutions that prepare ATMP-pnrb drugs retain the data in order to identify the tissue and cell samples used in the manufacture of the drugs (CSP art. R 5121-201-5, II). For their part, institutions that distribute or sell these drugs record a certain amount of data to enable the operation to be identified (CSP art. R.5121-201-5, III). These data are retained by the institutions for at least 30 years after the expiry date of the drug (CSP art. R 5121-201-7).

5.3.4. *Cell- or tissue-based therapy preparations*

Cell therapy preparations are not considered to be drugs if they do not involve substantial modifications during their manufacturing process such as to affect the properties of the initial biological identity and if their physiological function is the same in the donor and the recipient (for example, hematopoietic stem cells). Since they are not drugs, they follow the regime applied to cell products for therapeutic purposes (see CSP art. L 1243-1).

The administration of cell preparations may only be performed in health facilities, except in the case of derogations set by the Health Minister (CSP art. L1243-6).

* *

*

Stem cells are "carriers of immortality", to quote the Nicole Le Douarin's beautiful expression [LED 07], suggesting a field of research, development and considerable exploitation in fate and the future. For humans, this potential offers scientific resources to develop knowledge and therapeutic resources to provide higher performance medical care. However, this potential also presents a risk, if not a threat, for the human species because stem cells are at the origin of life, they multiply, specialize and aggregate to form a living body. Therefore, any scientific research and medical exploitation activity conflicts with the law, which considers the human person in their totality, the body being the necessary support of the person. As a consequence, the legal protection of the person includes upstream the essential protection of the human body. The three principles that ensure its protection are set out in Article 16-1 of the Civil Code: "*Every individual has the right to respect for their body. The human body is inviolable. The human body, its elements and its products cannot be the subject of a right of ownership*". Beyond the individual, however, it is the protection and preservation of the human species that is at stake.

Consequently, scientific and medical research finds itself between two principles that can be opposed, that of freedom of research and that of respect for human dignity. The balance between the two is difficult to maintain because, due to the need for evolution, it is necessary to impose limits on the freedom of research without hindering its pioneering, evolving and uncertain nature [PAR 09]. The fact that the law provides for a right of conscientious objection to the benefit of the researcher in the field of research into the embryo or embryonic stem cells is significant of the conflict between these two principles (CSP art. L 2151-7-1). The principle of safeguarding the dignity of the person recognized by the French Constitutional Council as a principle with constitutional value (Cons Const., Decision No. 94-343/344 DC of July 27, 1994) and its manifestation in other principles set by the Civil Code (art. 16 *et seq.*), in particular the principle of respect for the human being from the beginning of life, are markers of this balance for research into human stem cells (regarding these principles, see in particular [LEG 13e, BIN 15d]). Beyond these markers, the law can simply accompany the evolution of research as revealed by the successive revisions of bioethics law: 2004, then 2011 and 2018 to come (Law no. 2011-814 of July 7, 2011, art. 47-I).

With regard to the patentability of stem cell research and cell- and tissue-therapy medicinal products, the principle of freedom of trade proper to any market economy must be reconciled with the principle of non-ownership of the human body also set out in the Civil Code (arts. 16-5 and 16-6). The human body, its elements and its products cannot as such be the subject of a patent (CPI art. L 611-7), that is to say that the biotechnological invention does not benefit from the ownership protection granted to the inventor in the form of a temporary monopoly of exploitation. In other words, cell- and tissue-therapy medicinal products are subject to differentiated regulatory constraints on access to the drug market. It must be understood that these new drugs conduct a transformation of elements or products of the human body into health products, drugs that are set to become consumer goods governed by the rules of the market: profit, competition, advertising, etc. This transformation of elements or products of the human body into commercial health products weakens the foundations formed by the fundamental principles of human dignity and non-ownership of the human body *inter alia*.

The balance between respect for life from the beginning and therapies likely to improve the situation of patients can only evolve in favor of the latter within the tangential limit of the "attack on the integrity of the human species" (Civil Code, art. 16-4.).

The entire regulatory framework governing stem cells demonstrates the interference of the law in the field of health, which can be measured quantitatively by the yardstick of the imposing Public Health Code. This can also be observed at European level in the many health directives and regulations that have been adopted. More specifically, by introducing a statute for advanced-therapy medicinal products by regulation, the European Union intends to take advantage of the immediate harmonization effect of this instrument to foster innovation within its pharmaceutical policy by placing researchers in an emulation situation. At the same time, the Union pursues economic objectives by ensuring the proper functioning of the domestic market within the biotechnology sector and the competitiveness of the players in this sector [TAB 13]. As such, with a status proper to advanced-therapy medicinal products, the European Union is ahead of the United States in a highly competitive international sector.

Whilst the law in the field of healthcare can have a structuring role, bioethics law remains somewhat largely a "law of engineers" [STR 06].

Glossary

Adipose tissue-derived stem cells: adipose tissue-derived SCs can differentiate into adipocytes and many cell types, including osteocytes and chondrocytes. They possess immunosuppressive properties and low immunogenicity.

Adult stem cells: SCs present in an organ or a living tissue in minimal amounts. Their role is to maintain and repair the tissue where they are located (muscle, brain, blood, etc.). The term "somatic SCs" is sometimes used instead of "adult SCs", whereby "somatic" refers to the body and not to the germ cells, sperm or egg.

Advanced therapies: drug and therapeutic strategy produced from a biological source. Cell therapy and tissue engineering are part of innovative therapies and produce advanced therapy medicinal products.

Advanced therapy medicinal products (ATMP): set of medications formerly known as biological drugs, which include products derived from genetic therapy, somatic cell therapy and tissue and cell engineering.

Advanced therapy medicinal products prepared on a non-routine basis (ATMP-pnrb): the manufacture of which is authorized by the Member State of the European Union and specially designed for a given patient.

Bioethics: application of ethics, i.e. moral order, in the field of medicine, health and the study of life in general. Bioethics develops thinking on the ethical principles related to research in biology, medicine and genetics, and on their applications, with the aim of ensuring respect for human dignity.

Biomaterials: non-living substance or material implanted into the human organism to replace or repair a tissue. Biomaterials are used in regenerative medicine; they are biocompatible, of natural or synthetic origin.

Bioprinting: consists of the addition of cells, growth factors and other functional elements in superimposed layers. Through depositing the material and organization of the 3D structures, it is possible to produce three-dimensional matrices, or *3D matrices*, or even whole organs.

Bone marrow stem cells: bone marrow is a tissue found inside bones that constitutes the major source of SCs. It produces blood SCs and mesenchymal (non-hematopoietic) SCs.

Cardiac stem cells: SCs and progenitors distributed within the adult cardiac muscle (myocardium). Under normal conditions, cardiac progenitors are quiescent and do not contribute to the renewal of *cardiomyocytes*, specific muscle cells of the heart. After an injury or a heart attack, these progenitors are activated and differentiated into new myocytes and vascular cells.

Cardiotoxicity: the toxic effect of a substance on the heart or, more broadly, on the cardiovascular system. It is expressed by functional disorders of the cardiac muscle (myocardium), such as rhythm disorders, conduction disorders and high blood pressure.

Cell-based therapy: transplanting of cells aiming to differentiate into specific cells of a tissue or an organ in order to replace defective cells or restore the damaged tissue.

Cell therapy preparation (CTP): these preparations are not ATMPs because the changes made to the tissues or cells are not substantial, that is to say their properties are not modified (e.g. hematopoietic SCs for hematological reconstruction).

Clinical trials: scientific study evaluating the safety and efficacy of a drug in humans. Preceded by a pre-clinical phase that enables testing of the molecule on an animal or cellular model, clinical trials consist of three phases. Phase I evaluates the toxicity of the drug. Phase II determines the minimum effective dose of the drug and its side effects. Phase III evaluates the effectiveness of the drug on a larger cohort of patients, compared to a reference treatment or a placebo.

Differentiation: specialization of a cell line up to the ultimate stage of the specific functional cell of an organ. SCs have the ability to differentiate into one or more cell types that will form tissues and organs *in vivo*. This is the signature of SCs.

EMA (European Medicines Agency): European agency that evaluates drugs for human and veterinary use. The EMA's mission is to promote scientific excellence in the evaluation and supervision of medicines in the European Union.

Embryonic stem cells: SCs of the embryo a few days after fertilization. In humans, the embryo is not yet implanted (preimplantation embryo); it contains a compact mass of cells, which are the pluripotent embryonic SCs.

Epigenetics: regulation of gene activity that does not make use of DNA mutations. It corresponds to modifications (or factors) that are not encoded by the sequence of DNA (methylations) and that facilitate or prevent the expression of genes during cell division.

FDA (Food and Drug Administration): US food and drug agency, responsible for drug control and regulation. This government agency is divided into five centers that control food, drugs, animal food, dietary supplements, medical devices, and biological and blood products.

Hematopoietic stem cells: SCs at the origin of all blood cells. In adults, hematopoietic SCs generate one billion red blood cells and 100 million white blood cells an hour, as well as platelets and other mature blood cells.

Hepatotoxicity: the toxic effect of a substance on the liver. An organ that removes harmful substances from the body, the liver is exposed to chemicals and drugs that can cause inflammation (hepatitis) or necrosis of the liver cells.

High-throughput screening: high-throughput analyses used in pharmaceutical research. Potential drugs (drug candidates) are tested in miniaturized cell cultures with automated systems that enable the screening of hundreds of molecules or different culture conditions.

Induced pluripotent stem cells (IPS): pluripotent SCs obtained by introducing embryonic genes into somatic cells. These IPS cells are capable of self-renewal and pluripotency, two properties similar to those of embryonic SCs.

Intestinal stem cells: epithelial SCs of the intestine that ensure the extremely rapid renewal of the intestinal epithelium at the origin of different cell types of the intestinal epithelium.

Mesenchymal stem cells: SCs from bone marrow and from other tissues, adipose for example. They represent a heterogeneous population that has no common identity, no locating unit and no common future. They are able to differentiate into cartilage cells (chondrocytes), bone cells (osteoblasts) and fat cells (adipocytes).

Multipotency: ability of a cell to differentiate into several cell types, having the same embryonic origin.

Nanofiber: fiber less than one micron in size and approximately 50 nanometers (nm) in diameter. Nanofibers generate a large surface area to volume ratio and are used in many biomedical applications.

Nanotechnology: the technology that works at the scale of the molecule, between 1 and 100 nanometers (nanoscale). The nanometer (nm) represents *one thousandth of a micron*, that is, 10^{-9} meters, the distance between two atoms. The goal of this technology is to individually control atoms and molecules in order to create new functional structures. The architecture of the biomaterial designs a topography and a surface, which influence the future of the SCs.

Neural stem cells: SCs of the adult neural tissue, responsible for neurogenesis. Neural SCs give rise to new neurons and glial cells involved in learning and memory; this is the locus of *adult neurogenesis*, as opposed to embryonic neurogenesis.

Pluripotency: the ability of a cell to differentiate into any tissue of the organism, but the inability to give rise to a complete individual. Pluripotent cells give rise to the three embryonic compartments (endoderm, mesoderm and ectoderm). Each embryonic layer generates determined cells for each organ.

Precursor cell: fully engaged cell that differentiates into a specialized cell. The precursor is not capable of infinite self-renewal; it can be divided a limited number of times. Under differentiation conditions, the precursor, for example osteoblastic or hepatic, no longer divides. It transforms into the final cell, the osteocyte or the hepatocyte, which has stopped dividing.

Predictive toxicology: evaluation and prediction of the toxicity of a drug in pharmaceutical research. High-throughput analyses and modeling are new technologies that complete the toxicological test strategy.

Progenitor cell: stem cell committed to a cell line which, after several stages, will give rise to one (or more) differentiated cell(s) of this line. For example, in the line of hematopoietic SCs, two types of progenitor exist, myeloid and lymphoid, which will guide towards these two differentiation pathways. The progenitor is pluripotent, ensuring the creation of a progeny.

Regenerative medicine: innovative therapeutic treatment based on SCs aimed at differentiating into specific cells of a tissue or an organ in order to replace deficient cells or restore damaged tissue.

Self-renewal: infinite multiplication of a cell. The cell gives rise to two identical daughter cells, possessing the same properties as the mother cell. This is the signature of stem cells (SCs).

Skeletal muscle stem cells: the satellite cell is the SC at the origin of myogenic progenitors, then of muscular fibers. These multinucleate and contractile myofibers result from the fusion between the myogenic progenitors, called myoblasts, and the existing myofibers.

Skin stem cells: SCs that ensure homeostasis and regeneration of the skin. Several SC populations exist: epidermal SCs, dermal mesenchymal SCs, hair follicle SCs and endothelial SCs.

Somatic cells: cells making up the human organism that are not reproductive cells. Somatic cells are diploid, while germ cells are haploid.

Stem cells (SCs): original cells of the organism and all organized tissues. They can multiply to infinity and produce all of the cells of the body throughout life.

Tissue engineering: technologies for the transformation and organization of cells, tissues and biomaterials. The final product can regenerate, repair or replace human tissue in regenerative medicine.

Totipotency: cells that are fit for anything in the sense that they have the ability to become a complete individual. After the fusion of two human gametes, the fertilized egg (zygote) possesses all differentiation capacities. The cells can form an embryo, as well as the placenta.

Umbilical cord blood stem cells: SCs from the blood of the umbilical cord. Blood collected from the umbilical cord and placenta is rich in hematopoietic SCs and hematopoietic progenitors.

Unipotency: a cell's ability to differentiate into a single cell type. The cell is committed to a unique lineage that enables a specific cell type to be produced.

3D printing: consists of creating a three-dimensional object by superimposing thousands of layers of matter, on the basis of a numerical model. This process requires software, a 3D printer and materials.

Bibliography

[AGE 15] AGENCE DE LA BIOMEDECINE, Le rapport médical et scientifique du prélèvement et de la greffe en France, 2015.

[ALL 16] ALLIANCE FOR REGENERATIVE MEDECINE, Annual Data Report on gene and cellular therapies and the regenerative medicine sector, 2016.

[ARR 15] ARRIGHI N., MORATAL C., CLEMENT N. *et al.*, "Characterization of adipocytes derived from fibro/adipogenic progenitors resident in human skeletal muscle", *Cell Death and Disease*, p. e1733, 2015.

[BAC 12] BACACHE M., LEROYER A.-M., "Convention d'Oviedo", *R.T.D.Civ.*, p. 785, 2012.

[BAR 07] BARKER N., VAN ES J.H., KUIPERS J. *et al.*, "Identification of stem cells in small intestine and colon by marker gene Lgr5", *Nature*, vol. 449, no. 7165, pp. 1003–1007, 2007.

[BAR 09] BARKER N., RIDGWAY R.A., VAN ES J.H. *et al.*, "Crypt stem cells as the cells-of-origin of intestinal cancer", *Nature*, vol. 457, no. 7229, pp. 608–611, 2009.

[BAR 14] BARMADA S.J., SERIO A., ARJUN A. *et al.*, "Autophagy induction enhances TDP43 turnover and survival in neuronal ALS models", *Nature Chemical Biology,* vol. 10, no. 8, pp. 677–685, 2014.

[BEC 63] BECKER A.J., MCCULLOCH E.A, TILL J.E., "Cytological demonstration of the clonal nature of spleen colonies derived from transplanted mouse marrow cells", *Nature*, vol. 197, pp. 452–454, 1963.

[BEL 79] BELL E., IVARSSON B., MERRILL C., "Production of a tissue-like structure by contraction of collagen lattices by human fibroblasts of different proliferative potential *in vitro*", *Proceedings of the National Academy of Sciences USA*, vol. 76, no. 3, pp. 1274–1278, 1979.

[BEL 04] BELLIVIER F., "La loi no. 2004-800 du 6 août 2004 relative à la bioéthique", *RTDCiv*, p. 787, 2004.

[BER 97] BERTHOD F., DAMOUR O., "*In vitro* reconstructed skin models for wound coverage in deep burns", *British Journal of Dermatology*, vol. 136, no. 6, pp. 809–816, 1997.

[BET 13] BETSCHINGER J., NICHOLS J., DIETMANN S. *et al.*, "Exit from pluripotency is gated by intracellular redistribution of the bHLH transcription factor Tfe3", *Cell*, vol. 153, pp. 335–347, 2013.

[BIL 07] BILLON N., IANNARELLI P., MONTENEIRO M.C. *et al.*, "The generation of adipocytes by the neural crest", *Development*, vol. 134, no. 12, pp. 2283–2292, 2007.

[BIL 12] BILLON N., DANI C., "Developmental origins of the adipocyte lineage, new insights from genetics and genomics studies", *Stem Cell Review*, vol. 8, no. 1, pp. 55–66, 2012.

[BIN 02] BINET J.R., *Droit et progrès scientifique*, PUF, 2002.

[BIN 12] BINET J.R., "Ratification de la Convention d'Oviedo: la fin d'une longue attente", *JCPG*, p. 3, 2012.

[BIN 13] BINET J.R., "Recherches sur l'embryon: fin d'un symbole éthique et abandon d'une illusion thérapeutique; à propos de la loi nos 2013–715 du 6 août 2013", *JCPG*, p. 905, 2013.

[BIN 15a] BINET J.R., "Respect et protection du corps humain – Eléments et produits du corps humain", *Juris Classeur Civil Code*, Art. 16 to 16-14, fasc. 20, nos 6–7, 2015.

[BIN 15b] BINET J.R., "Respect et protection du corps humain – Eléments et produits du corps humain", *Juris Classeur Civil Code*, Art. 16 to 16-14, fasc. 20, nos 9–10, 2015.

[BIN 15c] BINET J.R., "Respect et protection du corps humain – Eléments et produits du corps humain," *Juris Classeur Civil Code*, Art. 16 to 16–14, fasc. 20, nos 15–16, 2015.

[BIN 15d] BINET J.R., "Respect et protection du corps humain. Eléments et produits du corps humains", *Juris Classeur Civil Code*, Art. 16 to 16–14, Fasc. 20, no. 70, 2015.

[BIN 17] BINET J.R., "Bioéthique", *Juris Classeur Synthèse*, 2017.

[BIO 13] BIOY X., RIAL-SEBAG E., "L'autorisation de la recherche sur l'embryon: évolution ou révolution", *A.J.D.A*, p. 2204, 2013.

[BOR 05] BORGETTO M., "La révision des lois bioéthiques", *RDSS*, Dalloz, p. 183, 2005.

[BOR 15] BOROVIAK T., LOOS R., LOMBARD P. *et al.*, "Lineage-specific profiling delineates the emergence and progression of naive pluripotency in mammalian embryogenesis", *Developmental Cell,* vol. 35, no. 3, pp. 366–382, 2015.

[BOU 12] BOULET M., "L'embryon humain saisi par le droit: quelle définition juridique pour quel statut. A propos de l'arrêt de la CJUE du 18 octobre 2011, Brüstle, C 34/10", *Rev. gén. de droit médical*, p. 133, 2012.

[BRI 12] BRINDLEY D., MASON C., "Human embryonic stem cell therapy in the post-Geron era", *Regenerative Medicine,* vol. 7, pp. 17–18, 2012.

[BUG 12] BUGANIM Y., FADDA D.A., CHENG A.W. *et al.*, "Single-cell expression analyses during cellular reprogramming reveal an early stochastic and a late hierarchic phase", *Cell*, vol. 150, no. 6, pp. 1209–1222, 2012.

[CHA 03] CHAMBERS I., COLBY D., ROBERTSON M. *et al.*, "Functional expression cloning of Nanog, a pluripotency sustaining factor in embryonic stem cells", *Cell*, vol. 113, no. 5, pp. 643–655, 2003.

[CHE 06] CHEMTOB M.C., *La brevetabilité des innovations biotechnologiques appliquées à l'homme*, 3rd ed., Tec & Doc, 2006.

[CHE 12] CHEMTOB-CONCE M.C., "La non brevetabilité des procédés de recherche destructifs d'embryons humains", *Prop. Industrielle*, no. 1, p. 27, 2012.

[CHE 74] CHEN H., LEBLOND C.P., "Origin, differentiation and renewal of the four main epithelial cell types in the mouse small intestine", *The American Journal of Anatomy,* vol. 141, pp. 461–479, 1974.

[CHE 11] CHENEY DE BEAUPRE A., "La révision de la loi relative à la bioéthique", *Dalloz*, p. 2217, 2011.

[CHI 14] CHIA N.Y., CHAN Y.S., FENG B. *et al.*, "A genome-wide RNAi screen reveals determinants of human embryonic stem cell identity", *Nature*, vol. 468, pp. 316–320, 2014.

[CHU 06] CHUA K.N., CHAI C., LEE P.C. *et al.*, "Surface-aminated electrospun nanofibers enhance adhesion and expansion of human umbilical cord blood hematopoietic stem/progenitor cells", *Biomaterials*, vol. 27, no. 36, pp. 6043–6051, 2006.

[COM 14] COMMITTEE FOR MEDICINAL PRODUCTS FOR HUMAN USE (CHMP), Holoclar, International non-proprietary name: *Ex vivo* expanded autologous human corneal epithelial cells containing stem cells, Procedure No. EMEA/H/C/002450/0000, Assessment report, EMA/25273/2015, 2014.

[COU 86] COULOMB B., SAIAG P., BELL E. *et al.*, "A new method for studying epidermalization *in vitro*", *British Journal of Dermatology*, vol. 114, no. 1, pp. 91–101, 1986.

[DAL 07] DALBY M.J., GADEGAAD N., TARE R. *et al.*, "The control of human mesenchymal cell differentiation using nanoscale symmetry and disorder", *Nature Biomaterials*, vol. 6, no. 12, pp. 997–1003, 2007.

[DAS 15] DAS S., PATI F., CHOI Y.J. *et al.*, "Bioprintable, cell-laden silk fibroin-gelatin hydrogel supporting multilineage differentiation of stem cells for fabrication of three-dimensional tissue constructs", *Acta Biomaterialia*, vol. 11, pp. 233–246, 2015.

[DAU 58] DAUSSET J., "Iso-leuco-anticorps", *Acta haematologica*, vol. 20, pp. 156–166, 1958.

[DIA 15] DIAS A.D., The influence of cellular patterning on brown adipogenic differentiation from mouse embryonic stem cells, PhD dissertation, Rensselaer Polytechnic Institute, New York, 2015.

[DIG 07] DI GIORGIO F.P., CARRASCO M.A., SIAO M.C. *et al.*, "Non-cell autonomous effect of glia on motor neurons in an embryonic stem cell-based ALS model", *Nature Neurosciences*, vol. 10, pp. 608–614, 2007.

[DUA 13] DUARTE CAMPOS D.F., BLAESER A., WEBER M. *et al.*, "Three-dimensional printing of stem cell-laden hydrogels submerged in a hydrophobic high-density fluid", *Biofabrication*, vol. 5, p. 015003, 2013.

[EVA 80] EVANS M.J., KAUFMAN M.H., "Establishment in culture of pluripotential cells from mouse embryo", *Nature*, vol. 292, pp. 154–166, 1980.

[FRA 06] FRAGNIEZ P.L., *Cellules souches et choix éthiques, Rapport au Premier ministre*, La Documentation française, 2006.

[FRI 76] FRIDENSTEIN A.J., GORSKAJA J.F., KULAGINA N.N., "Fibroblast precursors in normal and irradiated mouse hematopoietic organs", *Experimental Hematology*, vol. 4, no. 5, pp. 267–274, 1976.

[GAE 11] GAEBEL R., MA N., LIU J. *et al.*, "Patterning human stem cells and endothelial cells with laser printing for cardiac regeneration" *Biomaterials*, vol. 32, pp. 9218–9230, 2011.

[GAE 12] GAETANI R., DOEVENDANS P.A., METZ C.H. *et al.*, "Cardiac tissue engineering using tissue printing technology and human cardiac progenitor cells", *Biomaterials*, vol. 33, no. 6, pp. 1782–1790, 2012.

[GAF 13] GAFNI O., WEINBERGER L., MANSOUR A.A. *et al.*, "Derivation of novel human ground state naïve pluripotent stem cells", *Nature*, vol. 504, no. 7479, pp. 282–286, 2013.

[GAL 04a] GALLOUX J.C., "La loi du 6 août 2004 relative à la bioéthique", *R.T.D. Com.*, p. 707, 2004.

[GAL 04b] GALLOUX J.C., "Loi no. 2004–800 du 6 août 2004 sur la bioéthique", *Dalloz*, p. 2379, 2004.

[GAL 04c] GALLOUX J.C., GUTMAN E., "La protection des innovations biotechnologiques selon la loi du 6 août 2004: du génie génétique à la tératogénie", *Prop. Intellect.*, p. 871, 2004.

[GAL 04d] GALLOUX. J.C., "Les dispositions de la loi du 6 août 2004 relatives aux inventions biotechnologiques", *RDSS*, Dalloz, 206, 2004.

[GAL 08] GALLOUX J.C., GAUMONT-PRAT H., "Pan. Droit et libertés corporelles", *Dalloz*, p. 1435, 2008.

[GAL 09] GALLOUX J.C., "Non à l'embryon industriel: le droit européen des brevets au secours de la bioéthique?", *Dalloz*, p. 578, 2009.

[GAL 12a] GALLOUX J.C., "L'embryon communautaire", *Dalloz*, p. 410, 2012.

[GAL 12b] GALLOUX J.C., GAUMONT-PRAT H., "La non patrimonialité: la brevetabilité des embryons", *Dalloz*, p. 319, 2012.

[GAN 13] GANDHIMATHI C., VENUGOPAL J., RAVICHANDRAN R. *et al.*, "Mimicking nanofibrous hybrid bone substitute for mesenchymal stem cells differentiation into osteogenesis" *Macromolecular Bioscience*, vol. 13, no. 6, pp. 696–706, 2013.

[GEU 15] GEULA S., MOSHITCH-MOSHKOVITZ S., DOMINISSINI D. *et al.*, "m6A mRNA methylation facilitates resolution of naive pluripotency toward differentiation", *Science*, vol. 347, pp. 1002–1006, 2015.

[GIA 99] GIANCOTTI F.G., RUOSLATHI E., "Integrin signaling", *Science*, vol. 285, no. 5430, pp. 1028–1032, 1999.

[GIL 86] GILLERY P., MAQUART F.X., BOREL J.P., "Fibronectin dependence of the contraction of collagen lattices by human skin fibroblasts", *Experimental Cell Research*, vol. 167, no. 1, pp. 29–37, 1986.

[GLU 89] GLUCKMAN E., BROXMEYER H.A., AUERBACH A.D. *et al.*, "Hematopoietic reconstitution in a patient with Fanconi's anemia by means of umbilical-cord blood from an HLA-identical sibling", *New England Journal of Medicine*, vol. 321, no. 17, pp. 1174–1178, 1989.

[GRE 77] GREEN H., RHEINWALD J.G., SUN T.T., "Properties of epithelial cell type in culture: the epidermal keratinocyte and its dependence on products of the fibroblast", *Progress in Clinical and Biological Research*, vol. 17, pp. 493–500, 1977.

[GRI 94] GRINNELL F., "Fibroblasts, myofibroblasts and wound contraction", *Journal of Cellular Biology,* vol. 124, no. 4, pp. 401–404, 1994.

[GRU 11] GRUENE M., PFLAUM M., HESS C. *et al.*, "Laser printing of three-dimensional multicellular arrays for studies of cell–cell and cell–environment interactions", *Tissue Engineering. Part C Methods*, vol. 17, pp. 973–982, 2011.

[GUO 11] GUO L., ABRAMS R.M., BARBIAZ J.E. *et al.*, "Estimating the risk of drug-induced proarrhythmia using human induced pluripotent stem cell-derived cardiomyocytes", *Toxicological Sciences*, vol. 123, no. 1, pp. 281–289, 2011.

[GUR 62] GURDON J.B., "The developmental capacity of nuclei taken from intestinal epithelium cells of feeding tadpoles", *Journal of Embryology and Experimental Morphology*, vol 10, pp. 622–640, 1962.

[HAC 13] HACKETT J.A., DIETMANN S., MURAKAMI K. *et al.*, "Synergistic mechanisms of DNA demethylation during transition to ground-state pluripotency", *Stem Cell Report*, vol. 1, pp. 518–531, 2013.

[HAC 14] HACKETT J.A., SURANI M.A., "Regulatory principles of pluripotency: from the ground state up", *Cell Stem Cell*, vol. 15, no. 4, pp. 416–430, 2014.

[HAN 07a] HANNA J., CHENG A.W., SAHA K. *et al.*, "Human embryonic stem cells with biological and epigenetic characteristics similar to those of mouse ESCs", *Proceedings of the National Academy of Sciences USA*, vol. 107, no. 20, pp. 9222–9227, 2007.

[HAN 07b] HANNA J., WERNING M., MARKOULAKI S. *et al.*, "Treatment of sickle cell anemia mouse model with iPS cells generated from autologous skin", *Science*, vol. 318, no. 5858, pp. 1920–1923, 2007.

[HAN 09] HANNA J., MARKOULAKI S., MITALIPOVA M. *et al.*, "Metastable pluripotent states in NOD-mouse derived ESCs", *Cell Stem Cell*, vol. 4, no. 6, pp. 513–524, 2009.

[HAU 12] HAUSER J., "Brevetabilité des inventions sur l'embryon (suite et fin)", *R.T.D.Civ.*, p. 85, 2012.

[HAY 11] HAYASHI K., OHTA H., KURIMOTO K. *et al.*, "Reconstitution of the mouse germ cell specification pathway in culture by pluripotent stem cells", *Cell*, vol. 146, pp. 519–532, 2011.

[HEN 12] HENNET-VAUCHEZ S., "L'embryon de l'Union", *R.T.D. Européen*, p. 355, 2012.

[HOL 14] HOLMGREN G., SJOGREN A.K., BARRAGAN I. *et al.*, "Long-term chronic toxicity testing using human pluripotent stem cell-derived hepatocytes", *Drug Metabolism and Disposition*, vol. 42, no. 9, pp. 1401–1406, 2014.

[HUA 09] HUANG J., GRATER S.V., CORBELLINI F. *et al.*, "Impact of order and disorder in RGD nanopatterns on cell adhesion", *Nano Letters*, vol. 9, no. 3, pp. 1111–1116, 2009.

[HUA 14] HUANG K., WU Z., LIU Z. *et al.*, "Selective demethylation and altered gene expression are associated with ICF syndrome in human-induced pluripotent stem cells and mesenchymal stem cells", *Human Molecular Genetics*, vol. 23, no. 24, pp. 6448–6457, 2014.

[HUD 03] HUDON V., BERTHOD F., BLACK A.F. *et al.*, "A tissue-engineered endothelialized dermis to study the modulation of angiogenic and angiostatic molecules on capillary-like tube formation *in vitro*", *British Journal of Dermatology*, vol. 148, no. 6, pp. 1094–1104, 2003.

[IDE 09] IDELSON M., ALPER R., OBOLENSKY A. *et al.*, "Directed differentiation of human embryonic stem cells into functional retinal pigment epithelium cells", *Stem Cell Reports*, vol. 5, no. 4, pp. 396–408, 2009.

[JIN 11] JIN Z.B., OKAMOTO S., OSAKADA F. *et al.*, "Modeling retinal degeneration using patient-specific induced pluripotent stem cells", *PLoS One*, vol. 6, no. 2, p. e17084, 2011.

[KEI 05] KEIRSTEAD H.S., NISTOR G., BERNAL G. *et al.*, "Human embryonic stem cell-derived oligodendrocyte progenitor cell transplants remyelinate and restore locomotion after spinal cord injury", *Journal of Neurosciences*, vol. 25, pp. 4694–4705, 2005.

[KIJ 12] KIJENSKA E., PRABHAKARAN M.P., SWIESZKOWSKI W. *et al.*, "Electrospun bio-composite P(LLA-CL)/collagen I/collagen III scaffolds for nerve tissue engineering", *Journal of Biomedical Materials Research Part B: Applied Biomaterials*, vol. 100, no. 4, pp. 1093–1102, 2012.

[KIM 13] KIM H., WU J., YE S. *et al.*, "Modulation of β-catenin function maintains mouse epiblast stem cell and human embryonic stem cell self-renewal", *Nature Communications*, vol. 4, p. 2403, 2013.

[KOC 16] KOCH L., KUHN S., SORG H. *et al.*, "Laser printing of skin cells and human stem cells", *Tissue Engineering. Part C Methods*, vol. 16, pp. 847–854, 2016.

[KOJ 14] KOJIMA Y., KAUFMAN-FRANCIS K., STUDDERT J.B., "The transcriptional and functional properties of mouse epiblast stem cells resemble the anterior primitive streak", *Cell Stem Cells*, vol. 14, pp. 107–102, 2014.

[LAM 05] LAMBERT T., "Le brevet et la personne", *Dalloz*, p. 2005, 2005.

[LED 07] LE DOUARIN N., *Les cellules souches, porteuses d'immortalité*, Odile Jacob, 2007.

[LEE 09] LEE G., PAPAPETROU E.P., KIM H. *et al.*, "Modelling pathogenesis and treatment of familial dysautonomia using patient-specific iPSCs", *Nature*, vol. 461, no. 7262, pp. 402–406, 2009.

[LEE 12] LEE G., RAMIREZ C.N., KIM H. *et al.*, "Large-scale screening using familial dysautonomia induced pluripotent stem cells identifies compounds that rescue IKBKAP expression", *Nature Biotechnology*, vol. 30, no. 12, pp. 1244–1248, 2012.

[LEG 13a] LEGROS B., *Droit de la bioéthique*, no. 3, Les Etudes Hospitalières, 2013.

[LEG 13b] LEGROS B., *Droit de la bioéthique*, nos 8–15, Les Etudes Hospitalières, 2013.

[LEG 13c] LEGROS B., *Droit de la bioéthique*, nos 4–7, Les Etudes Hospitalières, 2013.

[LEG 13d] LEGROS B., *Droit de la bioéthique*, no. 20, Les Etudes Hospitalières, 2013.

[LEG 13e] LEGROS B., *Droit de la bioéthique*, no. 17, Les Etudes Hospitalières, 2013.

[LEN 91] LENOIR R., *Aux frontières de la vie: une éthique médicale à la française, Rapport au premier ministre*, La Documentation française, 1991

[LER 13] LEROYER A.M., "Chroniques", *R.T.D.Civ.*, p. 895, 2013.

[LIA 13] LIANG P., LAN F., LEE A.S. *et al.*, "Drug screening using a library of human induced pluripotent stem cell-derived cardiomyocytes reveals disease specific patterns of cardiotoxicity", *Circulation*, vol. 127, no. 16, pp. 1677–1691, 2013.

[LIU 12] LIU L., YUAN Q., SHI J. *et al.*, "Chemically-defined scaffolds created with electrospun synthetic nanofibers to maintain mouse embryonic stem cell culture under feeder-free conditions", *Biotechnology Letters*, vol. 34, no. 10, pp. 1951–1957, 2012.

[MA 08] MA K., CHAN C.K., LIAO S. *et al.*, "Electrospun nanofiber scaffolds for rapid and rich capture of bone marrow-derived hematopoietic stem cells", *Biomaterials*, vol. 29, no. 13, pp. 2096–2103, 2008.

[MAL 12] MALHERBE B., GALLOUX J.C., "L'arrêt Brüstle: de la régulation du marché à l'expression des valeurs", *Prop. Industrielle*, no. 9, p. 7, 2012.

[MAR 75] MARTIN G.R., EVANS M.J., "Differentiation of clonal lines of teratocarcinoma cells: formation of embryoid bodies *in vitro*", *Proceedings of the National Academy of Sciences USA*, vol. 72, pp. 1441–1445, 1975.

[MAR 12] MARTIAL-BRAZ N., BINET J.R., "Exclusion de la brevetabilité des embryons à des fins de recherche scientifique", *J.C.P. G*, p. 146, 2012.

[MAT 93] MATTEÏ J.F., *La vie en questions: pour une éthique biomédicale. Rapport au Premier ministre*, La Documentation française, 1993.

[MAT 09] MATHIEU B., *La bioéthique*, Dalloz, 2009.

[MCN 15] MCNEISH J., GARDNER J.P., WAIGNER B.J. *et al.*, "From dish to bedside: lessons learned while translating findings from a stem cell model of disease to a clinical trial", *Cell Stem Cell*, vol. 17, pp. 8–10, 2015.

[MEN 15] MENASCHÉ P., VANNEAUX V., HAGÈGE A. *et al.*, "Human embryonic stem cell-derived cardiac progenitors for severe heart failure treatment: first clinical case report", *European Heart Journal*, vol. 36, no. 30, pp. 2011–2017, 2015.

[MIL 13] MILLER J.D., GNAT Y.M., KISHINEVSKY S. *et al.*, "Human iPSC-based modeling of late-onset disease via progerin-induced aging", *Cell Stem Cell*, vol. 13, no. 6, pp. 691–705, 2013.

[MIR 11] MIRKOVIC A., "La recherche sur l'embryon stoppée à Luxembourg?", *A.J. fam.*, p. 518, 2011.

[MIR 12] MIRKOVIC A., *La ratification (enfin!) de la convention d'Oviedo sur les droits de l'homme et la biomédecine*, Dalloz, 2012.

[MIT 03] MITSUI K., TOKUZAWA Y., ITOH H. *et al.*, "The homeoprotein Nanog is required for maintenance of pluripotency in mouse epiblast and ES cells", *Cell*, vol. 113, no. 5, pp. 631–642, 2003.

[NAU 97] NAU J.Y., "Le Comité d'éthique admet l'utilisation de l'embryon à des fins médicales", *Le Monde*, May 1997.

[NIS 11] NISHIMURA K., SANO M., OHTAKA M. *et al.*, "Development of defective and persistent Sendai virus vector: a unique gene delivery/expression system ideal for cell reprogramming", *The Journal of Biochemical Chemistry*, vol. 286, no. 6, pp. 4760–4771, 2011.

[NIW 09] NIWA H., OGAWA K., SHIMOSATO D. *et al.*, "A parallel circuit of LIF signaling pathways maintains pluripotency of mouse ES cells", *Nature*, vol. 460, no. 7251, pp. 118–122, 2009.

[NOR 11] NORI S., OKADA Y., YASUDA A. *et al.*, "Grafted human-induced pluripotent stem-cell-derived neurospheres promote motor functional recovery after spinal cord injury", *Proceedings of the National Academy of Sciences USA*, vol. 108, no. 40, pp. 16825–16830, 2011.

[NUS 84] NUSGENS B., MERRILL C., LAPIERE C. *et al.*, "Collagen biosynthesis by cells in a tissue equivalent matrix *in vitro*", *Collagen and Related Research*, vol. 4, no. 5, pp. 351–363, 1984.

[OCO 16] O'CONNELL C.D., DI BELLA C., THOMPSON F. *et al.*, "Development of the Biopen: a handheld device for surgical printing of adipose stem cells at a chondral wound site", *Biofabrication*, vol. 8, no. 015019, 2016.

[OKI 07] OKITA K., ICHISAKA T., YAMANAKA S., "Generation of germ-line competent induced pluripotent stem cells", *Nature*, vol. 448, pp. 313–317, 2007.

[OLI 15] OLIVEIRA L.M.A, FALOMIR-LOCKHART L.J. *et al.*, "Elevated α-synuclein caused by *SNCA* gene triplication impairs neuronal differentiation and maturation in Parkinson's patient-derived induced pluripotent stem cells", *Cell Death and Disease*, vol. 6, no. 11, p. e1994, 2015.

[ORK 11] ORKIN S.H., HOCHEDLINGE K., "Chromatin connections to pluripotency and cellular reprogramming", *Cell*, vol. 145, pp. 835–850, 2011.

[PAR 07] PARK J., BAUER S., VON DER MARK K., SCHMUKI P., "Nanosize and vitality: TiO_2 nanotube diameter directs cell fate", *Nano Letters*, vol. 7, no. 6, pp. 1686–1691, 2007.

[PAR 09] PARICARD S., "La recherche médicale et le droit: une relation ambivalente", *RDSS*, p. 98, 2009.

[PAU 05] PAUVERT B., PEDROT P., "La législation applicable aux cellules souches embryonnaires", *L.P.A.*, p. 77, 18 February 2005.

[PEI 08] PEIGNE J., "Le droit des biothérapies: entre subsidiarité éthique et harmonisation technique", *RDSS*, p. 292, 2008.

[PHI 94] PHILLIPS C.L., COMBS S.B., PINNELL S.R., "Effects of ascorbic acid on proliferation and collagen synthesis in relation to the donor age of human dermal fibroblasts", *Journal of Investigative Dermatology*, vol. 103, no. 2, pp. 228–232, 1994.

[POT 71] POTTER V.R., *Bioethics: Bridge to the Future*, Prentice Hall, Englewood Cliffs, 1971.

[RAM 13a] RAMPICHOVA M., CHVOJKA J., BUZGO M. *et al.*, "Elastic three-dimensional poly (ε-caprolactone) nanofibre scaffold enhances migration, proliferation and osteogenic differentiation of mesenchymal stem cells", *Cell Proliferation*, vol. 46, no. 1, pp. 23–37, 2013.

[RAM 13b] RAMSDEN C.M., POWNER M.B., CARR A.J. *et al.*, "Stem cells in retinal regeneration: past, present and future", *Development*, vol. 140, no. 12, pp. 2576–2585, 2013.

[RAT 90] RATHJEN P.D., TOTH S., WILLIS A. *et al.*, "Differentiation inhibiting activity is produced in matrix-associated and diffusible forms that are generated by alternate promoter usage", *Cell*, vol. 62, no. 6, pp. 1105–1114, 1990.

[RÉG 97] RÉGNIER M., STAQUET M.J., SCHMITT D. *et al.*, "Integration of Langerhans cells into a pigmented reconstructed human epidermidis", *Journal of Investigative Dermatology*, vol. 109, no. 4, pp. 510–512, 1997.

[ROS 90] ROSDY M., CLAUSS L.C., "Terminal epidermal differentiation of human keratinocytes grown in chemically defined medium on inert filter substrates at the air-liquid interface", *Journal of Investigative Dermatology*, vol. 95, no. 4, pp. 409–414, 1990.

[RUB 95] RUBIN J.S., BOTTARO D.P., CHEDID M. *et al.*, "Keratinocyte growth factor as a cytokine that mediates mesenchymal-epithelial interaction", *Experientia*, vol. 74, no. 8, pp. 191–214, 1995.

[SCH 08] SCHENKE-LAYLAND K., RHODES K.E., ANGELIS E. *et al.*,"Reprogrammed mouse fibroblasts differentiate into cells of the cardiovascular and hematopoietic lineages", *Stem Cells*, vol. 26, pp. 1537–1546, 2008.

[SCH 15] SCHWARTZ S.D., REGILLO C.D., LAM B.L. *et al.*, "Human embryonic stem cell-derived retinal pigment epithelium in patients with age-related macular degeneration and Stargardt's macular dystrophy: follow-up of two open-label phase 1/2 studies", *Lancet*, vol. 395, no. 9967, pp. 509–516, 2015.

[SHA 90] SHAHABEDDIN L., BERTHOD F., DAMOUR O. *et al.*, "Characterization of skin reconstructed on a chitosan-cross-linked collagen-glycosaminoglycan matrix", *Skin Pharmacology*, vol. 3, no. 2, pp. 107–114, 1990.

[SHI 13] SHIM J.H., LEE T.R., SHIN D.W., "Novel *in vitro* culture condition improves the stemness of human dermal stem/progenitor cells", *Molecules and Cells*, vol. 36, no. 6, pp. 556–563, 2013.

[SMI 10] SMITH A., "The Pluripotent stem cells: private obsession and public expectation", *EMBO Molecular Medicine*, vol. 2, pp. 113–116, 2010.

[SMI 88] SMITH A.G., HEATH J.K., DONALDSON D.D. *et al.*, "Inhibition of pluripotential embryonic stem cell differentiation by purified polypeptides", *Nature,* vol. 336, no. 6200, pp. 688–690, 1988.

[SON 15] SONG W.K., PARK K.M., KIM H.J. *et al.*, "Treatment of macular degeneration using embryonic stem cell-derived retinal pigment epithelium: preliminary results in Asian patients", *Stem Cell Reports*, vol. 4, no. 5, pp. 860–872, 2015.

[SOR 16] SORIN J., "Recherches sur des cellules souches: consentement du couple donneur", *A.J.D.A.*, p. 843, 2016.

[SOU 12] SOUFI A., DONAHUE G., ZARET K.S., "Facilitators and impediments of the pluripotency reprogramming factors' initial engagement with the genome", *Cell*, vol. 151, pp. 994–1004, 2012.

[SRE 13] SREEREKHA P.R., MENON D., NAIR S.V. *et al.*, "Fabrication of electrospun poly (lactide-co-glycolide)-fibrin multiscale scaffold for myocardial regeneration in vitro", *Tissue Engineering Part A*, vol. 19, nos 7–8, pp. 849–859, 2013.

[STR 06] STRICKLER Y., *Les biens*, PUF, 2006.

[TAB 13] TABOULET F., "Les médicaments de thérapie innovante: quelle spécificité en droit pharmaceutique?", *Quaderni*, no. 81, June 2013.

[TAK 06] TAKAHASHI K., YAMANAKA S., "Induction of pluripotent stem cells from mouse embryonic and adult fibroblast cultures by defined factors", *Cell*, vol. 126, pp. 663–676, 2006.

[TAK 07] TAKAHASHI K., TANABE K., OHNUKI M. *et al.*, "Induction of pluripotent stem cells from adult human fibroblasts by defined factors", *Cell*, vol. 131, no. 5, pp. 861–872, 2007.

[TES 07] TESAR P.J., CHENOWETH J.G., BROOK F.A. *et al.*, "New cell lines from mouse epiblast share defining features with human embryonic stem cells", *Nature*, vol. 448, no. 7150, pp. 196–199, 2007.

[THI 12] THIERRY D., "La France enfin liée par la Convention d'Oviedo sur les droits de l'homme et la biomédecine", *RDSS*, p. 839, 2012.

[THO 94] THOUVENIN D., "De l'éthique biomédicale aux lois "bioéthiques"", *R.T.D. Civ.*, p. 717, 1994.

[THO 98] THOMSON J.A., ITSKOVITZ-ELDOR J., SHAPIRO S.S. *et al.*, "Embryonic stem cells derived from human blastocysts", *Science*, vol. 282, pp. 1145–1147, 1998.

[THO 05] THOUVENIN D., *La loi relative à la bioéthique ou comment accroître l'accès aux éléments d'origine humaine*, Dalloz, 2005.

[TIL 61] TILL J.E., MCCULLOCH E.A., "A direct measurement of the radiation sensitivity of normal mouse bone marrow cells", *Radiation Research*, vol. 14, pp. 213–222, 1961.

[TIN 91] TINOIS E., TIOLLIER J., GAUCHERAND M. *et al.*, "*In vitro* and post-transplantation differentiation of human keratinocytes grown on the human type IV collagen film of a bilayered dermal substitute", *Experimental Cell Research*, vol. 193, no. 2, pp. 310–319, 1991.

[TOD 93] TODD C., HEWITT S.D., KEMPENAAR J. *et al.*, "Co-culture of human melanocytes and keratinocytes in a skin equivalent model: effect of ultraviolet radiation", *Archives of Dermatologic Research*, vol. 285, no. 8, pp. 455–459, 1993.

[TRE 05] TREMBLAY P.L., BERTHOD F., GERMAIN L. *et al.*, "*In vitro* evaluation of the angiostatic potential of drugs using endotheliaized tissue-engineerined connective tissue", *Journal of Pharmacology and Experimental Therapeutics*, vol. 315, no. 2, pp. 510 516, 2005.

[TUR 16] TURNER L., KNOEPFLER P., "Selling stem cells in the USA: assessing the direct-to-consumer industry", *Cell Stem Cell*, vol. 19, no. 2, pp. 154–157, 2016.

[TYL 16] TYL L., CHONG J.J.H., "Cardiac progenitor cells for heart repair", *Cell Death and Discovery*, vol. 2, no. 16052, 2016.

[VIG 11] VIGNEAU D., *Les dispositions de la loi 'bioéthique' du 7 juillet 2011 relatives à l'embryon et au fœtus humain*, Dalloz, 2011.

[VIN 12] VINES J.B., LIM D.J., ANDERSON J.M. *et al.*, "Hydroxyapatite nanoparticle reinforced peptide amphiphile nanomatrix enhances the osteogenic differentiation of mesenchymal stem cells by compositional ratios", *Acta Biomaterialia*, vol. 8, no. 11, pp. 4053–4063, 2012.

[WAD 57] WADDINGTON C.H., *The Strategy of the Genes. A Discussion of Some Aspects of Theoretical Biology*, George Allen & Unwin, 1957.

[WAN 10] WANG S., LU B., GIRMAN S. *et al.*, "Non-invasive stem cell therapy in a rat model for retinal degeneration and vascular pathology", *PLoS One*, vol. 5, no. 2, p. e9200, 2010.

[WAN 12] WANG P.Y., THISSEN H., TSAI W.B., "The roles of RGD and grooved topography in the adhesion, morphology, and differentiation of C2C12 skeletal myoblasts", *Biotechnology and Bioengineering*, vol. 109, no. 8, pp. 2104–2115, 2012.

[WAN 16] WANKHADE U.D., SHEN M., KOLHE R. *et al.*, "Advances in adipose-derived stem cells isolation, characterization, and application in regenerative tissue engineering", *Stem Cells International*, vol. 2016, no. 3206807, 2016.

[WAR 10] WARREN L., MANOS P.D., AHFELD T. *et al.*, "Highly efficient reprogramming to pluripotency and directed differentiation of human cells with synthetic modified mRNA", *Cell Stem Cell*, vol. 7, pp. 618–630, 2010.

[WIL 97] WILMUT I., SCHIEKE A., MCWHIR J. *et al.*, "Viable offspring derived from fetal and adult mammalian cells", *Nature*, vol. 385, pp. 810–813, 1997.

[WIL 07] WILKE R.A., LIN D.W., RODEN D.M. *et al.*,"Identifying genetic risk factors for serious adverse drug reactions: current progress and challenges", *Nature Reviews Drug Discovery*, vol. 6, no. 11, pp. 904–916, 2007.

[WU 15] WU J., OKAMURA D., LI M. *et al.*, "An alternative pluripotent state confers interspecies chimaeric competency", *Nature*, vol. 521, pp. 316–321, 2015.

[YAM 10] YAMAZAKI C.M., KADOYA Y., HOZUMI K. *et al.*, "A collagen-mimetic triple helical supramolecule that evokes integrin-dependent cell responses", *Biomaterials*, vol. 31, no. 7, pp. 1925–1934, 2010.

[YAN 05] YANG F., MURUGAN R., WANG S. *et al.*, "Electrospinning of nano/micro scale poly(L-lactic acid) aligned fibers and their potential in neural tissue engineering", *Biomaterials*, vol. 26, no. 15, pp. 2603–2610, 2005.

[YIN 03] YING Q.L., NICHOLS J., CHAMBERS I. *et al.*, "BMP induction of Id proteins suppresses differentiation and sustains embryonic stem cell self-renewal in collaboration with STAT3", *Cell*, vol. 115, pp. 281–292, 2003.

[YIN 08] YING Q.L., WRAY J., NICHOLS J. *et al.*, "The ground state of embryonic stem cell self-renewal", *Nature*, vol. 453, no. 7194, pp. 519–523, 2008.

[YIN 13] YIN H., PRICE F., RUDNICKI M.A., "Satellite cells and the muscle stem cell niche", *Physiological Reviews*, vol. 93, no. 1, pp. 23–67, 2013.

[YOU 15] YOUNG J.E., BOULANGER-WEILL J. *et al.*, "Elucidating molecular phenotypes caused by the SORL1 Alzheimer's disease genetic risk factor using human induced pluripotent stem cells", *Cell Stem Cell*, vol. 16, no. 4, pp. 373–385, 2015.

[YU 07] YU J., VODYANIK M.A., SMUGA-OTTO K. *et al.*, "Induced pluripotent stem cell lines derived from human somatic cells", *Science*, vol. 318, no. 5858, pp. 1917–1920, 2007.

[ZHA 15] ZHANG J., JIAO J., "Molecular biomarkers for embryonic and adult neural stem cell and neurogenesis", *BioMed Research International*, vol. 2015, no. 727542, 2015.

Index

T, U, Z

FSC
www.fsc.org
MIX
Papier | Fördert
gute Waldnutzung
FSC® C083411

Kolibri 360 GmbH
Ferdinand-Jühlke-Straße 7
99095 Erfurt, Deutschland
produktsicherheit@kolibri360.de